Future Flight

Future Flight

How Talent, Capital and Manufacturing will Shape the Next Era of Uncrewed Aviation

Sharon Rossmark

Zayos

A division of Zayos Global Ventures, LLC

Published by Zayos, a division of Zayos Global Ventures, LLC

Glenview, Illinois

ISBN: 978-1-948251-14-3

Library of Congress Control Number: 2026916540

FutureFlightBook.com

First edition, July 2026

1 3 5 7 9 10 8 6 4 2

Printed in the United States of America

Disclaimer: While the author has drawn on extensive professional experience in the uncrewed aviation industry, this book is intended as a general information and reference resource only. The author and publisher make no warranty or representation, express or implied, regarding the accuracy, completeness, or applicability of the material contained herein. Neither the author nor the publisher shall be liable for any direct, indirect, special, incidental, or consequential damages arising from the use of, or reliance on, any information, strategy, or recommendation contained in this book. Readers are encouraged to seek qualified professional advice before making decisions based on any material herein.

To the educators who create pathways that once seemed impossible.

To the students who step into this new era with curiosity and courage.

To the technicians, pilots, engineers, and operators whose skill turns innovation into reality.

To the community leaders who recognize opportunity where others see uncertainty.

To the policymakers who understand that the sky is both a shared resource and a shared responsibility.

To the investors who choose to fund the infrastructure, not just the headlines.

To the manufacturers who prove each day that American ingenuity is not a relic, it is a renewable resource.

And to every woman and girl who has ever looked up at the sky and wondered whether there was a place for her in it.

There is.

This book is for you.

Foreword

When I led the first deployment of small uncrewed aerial systems to a major disaster — Hurricane Katrina in 2005 — the technology performed surprisingly well. The drones flew, collected critical imagery, and delivered value. What followed was a wave of interest from the Federal Aviation Administration (FAA), the Association for Uncrewed Vehicle Systems International (AUVSI), the International Association of Emergency Managers, and other government and corporate stakeholders. Workshops, keynotes, and regulatory discussions proliferated. Everyone assumed fire and rescue teams would soon be using them routinely.

Yet adoption never came.

Engineers continued to improve flight controls, communications, and reliability, but the pattern repeated. After the 2011 Fukushima Daiichi nuclear disaster, drones performed even better and again generated praise and excitement — but still no widespread adoption. By the time Hurricane Harvey struck in 2017, drones had become dramatically cheaper and smarter, thanks to advances in sensors, mapping software, and artificial intelligence. They delivered impressive results and won awards. Once more, the operational world did not change.

Only in 2022 did we finally see a coordinated, strategic deployment by a state agency — Florida's State Emergency Response Team during Hurricane Ian. Even today, the use of uncrewed aviation in disaster management remains largely ad hoc.

The problem was never the drones. They did what they were designed to do. The real obstacles were the responders who had never trained with them, the incident command structures that didn't know how to integrate them, and the lack of capacity to receive, process, and act on the new streams of data. We had built capable machines and expected an unprepared world to spontaneously rewrite its workflows, update its acquisition processes, scale workforce development, and create coherent policy frameworks.

This same cycle — impressive technology followed by disappointing adoption — has played out across many applications of drones over the past two decades. The persistent gap is not a technology problem. It is a human systems problem. And it remains one of the most consequential unsolved challenges in robotics and autonomous systems.

We need to close this gap now. Not only to finally capture the opportunities we continue to miss in public safety, precision agriculture, infrastructure inspection, and logistics, but also because we are falling behind other nations in preparing for the profound changes already underway in aviation and artificial intelligence.

Future Flight is the book that anyone serious about the success of uncrewed aviation should read. Sharon Rossmark is uniquely qualified to write it. She is not an academic or a think-tank analyst. She brings more than a decade of direct, hands-on engagement with the people actually building the commercial drone ecosystem — operators, educators, investors, and policymakers. Before becoming a Part 107 pilot, she built a career as a senior corporate executive, giving her deep insight into how businesses adopt (or fail to adopt) disruptive technologies and develop the workforces they require.

Throughout her career, Sharon has been what Malcolm Gladwell would call a "connector" — someone exceptionally skilled at bringing the right people together. She founded and leads Women and Drones, an organization that champions greater participation in uncrewed aviation. She also excels at connecting ideas and expressing them without jargon or handwaving. Her combined experience in business, drones, workforce development, and investing allows her to both identify real problems and offer practical, grounded solutions.*

Read Future Flight. Then act on it.

Robin R. Murphy, Ph.D.
Raytheon Professor Emeritus of Computer Science and Engineering
Texas A&M University
College Station, Texas
June 2026.

Why This Book, Why Now

This book is an attempt to name that problem with precision and to propose a path forward with specificity. It is addressed to the executives building drone companies, the policymakers writing the rules, the educators building the pipelines, the investors allocating capital, and the community leaders making economic development decisions that will shape their regions for a generation. It is addressed, above all, to the people who understand that the window is open and that the answer to whether America will lead the uncrewed aviation era is being written right now, by the decisions being made, and not made, in this specific season of the industry's development.

The book is organized in four parts. Part I maps the landscape: the scale and velocity of the drone revolution, the geopolitical competition with China, the talent deficit that threatens to become the binding constraint on growth, and the manufacturing imperative that federal policy has made urgently real. Part II prescribes the workforce strategies: the education and credentialing pipeline, the business case for the full talent pool, the geographic communities the industry has not yet reached, and the employer strategies that must retain what the pipeline produces. Part III follows the money: public investment, private capital, defense spending, and the regional economic development competition that will determine where the industry's long-term value lands. Part IV looks toward the future: Advanced Air Mobility, the regulatory architecture the next decade requires, the integrated national strategy that workforce, investment, and manufacturing together demand, and the direct call to action that the preceding chapters build toward.

The argument that runs through all four parts is a single one: the United States has the technology, the capital, the talent base, and the regulatory foundations to lead the world in uncrewed aviation for the next generation. What it does not yet have is the coordinated national will to build what that leadership requires. That will is a choice, made by specific people in specific roles at a defined point in time. The point in time is now. The people making the choice are identifiable. This book is, among other things, addressed to them.

Acknowledgments

This book grew out of more than a decade of conversations in classrooms, at conferences, on university campuses, at summer camps, and on flight projects with the people who are building this industry from the ground up. The community behind this work is larger than any list I could write, and I want to acknowledge that plainly at the start.

To the educators and instructors who built UAS programs at colleges and universities across the country, many beginning with a single drone, a borrowed classroom, and the conviction that their students deserved a pathway into a growing field: your work is the foundation beneath every chapter. You built pipelines before the industry knew it needed them.

To the training organizations, from the online platforms that opened the door to Part 107 for hundreds of thousands of new pilots to the specialized providers who taught public safety agencies to fly with discipline and rigor, and the simulation companies that proved competence can be built before anyone touches an aircraft: thank you for showing that access and quality can coexist.

To the K–12 educators, youth program leaders, and STEM advocates who understood that the talent pipeline must begin long before college: the full return on your work will not be visible for years, but its impact is already shaping the future. I hope this book strengthens your case with the funders and policymakers who need to hear it.

To the global training community across regions and regulatory systems around the world, each building a drone workforce that reflects your own communities' needs: the argument of this book is international even when its policy lens is domestic. Your work matters, and it is seen.

To the operators, pilots, technicians, fleet managers, and data professionals who opened your doors, answered my questions, and told me the truth about what is working and what is not: the credibility of this book rests on your candor. I am grateful for every conversation.

To the investors and industry leaders who spoke honestly about capital allocation, beyond the press release version of this industry: the investment misalignment chapter exists because you trusted me with the real picture.

To the manufacturers, engineers, assembly teams, supply chain managers, and founders who are doing the patient, unglamorous work of building a domestic drone industrial base, component by component: your story is the one I most wanted to tell accurately. I hope I have done it justice.

To the Women and Drones community: everything I understand about the gap between this industry's stated values and its actual pipeline practices, I learned alongside you. This book is the argument your experience demands.

To my editor, whose judgment strengthened every chapter and whose patience with the process was a gift I did not always deserve.

To my family: thank you for the time, the space, and the belief that this was worth finishing.

I have spent more than ten years flying, teaching, speaking, and working alongside the people who are building this industry in real time. I have watched it grow not from a distance, but from the field, the classroom, the conference stage, and the flight line. The people in these pages, named and unnamed, are the ones who shaped that journey. This book exists because of them, and it is dedicated to their work

A Note on Perspective

This book was written from the inside. I have spent more than a decade working in the uncrewed aviation industry, building workforce programs, advising companies, attending the conferences, reading the rulemaking dockets, and sitting in the rooms where hiring decisions are made and strategic bets are placed. The arguments in this book are not assembled from a distance. They are drawn from that experience, tested against the evidence, and offered with the conviction of someone who has watched an extraordinary industry limit its own potential by failing to take its most solvable problems seriously.

The primary problems this book identifies are workforce, investment, and manufacturing. They are also, I would argue, three expressions of the same underlying problem: the gap between what the American uncrewed aviation industry is capable of building and what it has so far been willing to commit to building. That gap is not technological. The technology has never been the hard part. The hard part has always been developing the people, the systems, and the will to deploy it at scale. That sentence opens Chapter 1 for a reason. It is the book's thesis, stated plainly before the first scene unfolds.

Table of Contents

EPILOGUE

PART I: THE LANDSCAPE

Future Flight

How Talent, Capital and Manufacturing will Shape the Next Era of Uncrewed Aviation

About Part I

The four chapters of Part I establish the "state of play" — the technological momentum, geopolitical stakes, workforce shortfall, and manufacturing reality that every subsequent chapter builds upon. Before the book can prescribe solutions, it must earn the reader's understanding of the full scope of the challenge. Part I does that work: it situates the uncrewed aviation revolution in its proper context — not as a future possibility, but as a present reality demanding urgent, coordinated response. Readers who finish Part I will understand precisely why workforce, investment, and manufacturing sovereignty are not peripheral concerns. They are the whole game.

Cleared for Takeoff: The Uncrewed Aviation Revolution

"The technology was never the hard part. The hard part has always been developing the people, the systems, and the will to deploy it at scale."

— Sharon Rossmark

One Day in American Airspace

The sun has not yet risen over Assumption, Illinois, when the first drone of the day lifts off. It climbs smoothly into the pre-dawn dark, its rotors a low hum against the silence of the cornfields stretching to the horizon. The operator, working from a truck parked at the edge of a gravel road, watches the flight path populate on a tablet screen. What she is watching is not a test. It is Tuesday morning, and 200 acres of corn need spraying before the heat of the day sets in.

The drone, an agricultural unmanned aircraft system (UAS) equipped with GPS navigation and a multispectral imaging array, maps crop health, soil moisture, and pest pressure in real time, covering the field in a fraction of the time a ground-based sprayer would require, with centimeter-level precision that no tractor operator, however skilled, could match. By the time the sun clears the tree line, the field is done. The drone is back on its landing pad. The data it collected is already uploading to a cloud platform where an agronomist three states away will review it before lunch.

By mid-morning, a utility corridor northeast of Pittsburgh is receiving an inspection it could not have had without drones. Two years ago, walking that 138,000-volt transmission-line right-of-way required a contracted helicopter, a two-person crew, and a full day of coordination, logistics, and post-flight review that cost the utility company tens of thousands of dollars per inspection cycle. Today, a single licensed remote pilot guides a UAS through the corridor, capturing 4K thermal imagery of every tower, connector, and insulator stack, flagging potential fault points before they become outages. The helicopter that once did this job is grounded. Not because it failed. Because it is no longer the best tool for this particular mission.

In the early afternoon, a suburban neighborhood in Phoenix sees something that would have seemed implausible a decade ago: a small uncrewed aircraft descends from cruising altitude, hovers above a backyard at a designated delivery point, and lowers a package on a tether before releasing it and climbing away. Inside the package is a prescription medication. The homeowner, a woman in her seventies who does not drive, tracked the delivery on her phone with the same casual familiarity she applies to tracking a pizza order. She did not think about the airspace management software, the remote identification broadcast, or the regulatory framework that made this delivery legal. She simply knew her medication was coming. Then it arrived. Technology had become, in the most important sense, invisible.

Forty miles offshore in the Gulf of Mexico, a long-endurance fixed-wing drone sweeps a pipeline corridor in the early afternoon. There is no crew out here. No boat. No helicopter thundering above the water. There is a sensor platform transmitting high-resolution pipeline data in real time to an operations center onshore, where engineers monitor it on large display screens. The drone covers the corridor in hours. It will land, recharge, and be ready for the next mission by morning. The offshore energy company that operates it has calculated the savings: fewer vessel deployments, fewer personnel exposure hours, fewer days of disruption to the platform operations below. The math makes the drone not just useful, but essential.

As the sun sets over far west Texas, a different kind of mission is underway along the U.S.–Mexico border. A Customs and Border Protection (CBP) drone has been on station since early afternoon, sweeping a 60-mile corridor in a persistent surveillance pattern that no single crewed aircraft could sustain without crew rotation and refueling stops. The imagery it collects feeds directly to ground agents. It sees terrain that vehicles cannot reach and stays on station longer than any pilot can safely remain airborne. A decade ago, this mission required multiple aircraft and rotating crews. Tonight, it requires one drone and a remote operator working from a climate-controlled ground control station.

Pull back far enough from these five scenes, the Illinois field, the Pittsburgh power line, the Phoenix doorstep, the Gulf of Mexico, the Texas borderland and something important becomes visible that each individual scene obscures. This is not a technology demonstration. This is not a pilot program or a proof of concept being evaluated for possible future deployment. This is Tuesday. This is the ordinary, unremarkable present tense of an industry that has moved from novelty to infrastructure with a speed that has left policy, education, and public perception struggling to keep pace. And it is, by every available measure, only the beginning.

Precision agriculture drones are among the fastest-growing commercial UAS applications globally, transforming field operations that once required hours of ground-level labor.

The Numbers That Clarify

Numbers have a way of either clarifying or obscuring, depending on how they are presented. The numbers behind the uncrewed aviation revolution are genuinely clarifying, not because they are large, though they are, but because of what they reveal about the structure and trajectory of an industry in the middle of a steep and accelerating growth curve.

The global UAV market was valued at $26.12 billion in 2025, according to MarketsandMarkets, and is projected to reach $40.56 billion by 2030, compounding at a rate of 9.2 percent annually.[1] Those figures capture hardware, software, services, and the expanding ecosystem of data products that drones generate. The commercial drone installed base worldwide reached 2.8 million units in 2024 and is on a trajectory to reach 4.5 million by 2029.[2] These are not aspirational numbers. They are built from actual fleet counts, sales data, and deployment records across industries that have already made the operational and financial commitment to uncrewed aviation.

By 2025, active Part 107 Remote Pilot Certificates had approached 492,000, closing the gap with all non-student traditional pilots combined, a demographic shift that has unfolded with almost no public notice.

The human dimension of this growth is equally striking. According to the Federal Aviation Administration's (FAA) Civil Airmen Statistics for December 31, 2025, there are 492,311 active Part 107 Remote Pilot certificates in the United States, approaching the total count of all non-student traditional pilots (private, commercial, and airline transport) combined, a gap that has narrowed from hundreds of thousands just a decade ago.[3] Nearly 74,000 aspiring drone pilots sat for the Part 107 knowledge test in 2025 alone, a figure close to double the test volume from 2020, according to FAA testing data.[10]

Globally, the workforce picture is similarly expansive. More than 2.1 million people worldwide now work in the drone industry, across roles ranging from hardware manufacturing and software development to operations, data analysis, regulatory compliance, and fleet management.[4] Investment flows confirm the scale of market confidence: the drone sector has attracted more than 7,000 funding rounds from over 2,000 investors, with an average deal size of $27.2 million per round.[4] That is not venture capital dabbling in an interesting technology. That is the allocation of serious capital toward what investors perceive as durable, structural growth.

The fastest-growing segment within the UAV market, according to MarketsandMarkets, is fully autonomous operations: systems that execute missions without continuous human input, guided by artificial intelligence (AI), edge computing, and advanced sensor fusion.[1] This is where the economics become transformational. Autonomy is the lever that converts a drone from a remotely operated tool into a scalable, deployable asset that can operate continuously across multiple sites with minimal human intervention per flight hour.

These numbers invite comparison to other technologies at similar inflection points. The commercial internet in 1997 was growing at double-digit rates annually, but its full economic impact was invisible to most observers for another decade. The smartphone ecosystem in 2008 was already generating extraordinary investment and user adoption, but the app economy that would eventually reshape retail, transportation, media, and communication was still largely unbuilt. Commercial aviation in the early 1950s was technically capable but operationally constrained, waiting for jet engines, regulatory frameworks, and airport infrastructure to unlock mass-market travel. In each case, the inflection point was visible in retrospect. The question was always whether those who recognized it early had the institutional capacity to build what the moment required.

What distinguishes uncrewed aviation from most previous technology revolutions is the breadth of its application base. The internet replaced one information infrastructure. The smartphone consolidated communication and computing into a single device. Drones do not replace one industry. They cut across all of them simultaneously agriculture, energy, logistics, construction, public safety, defense, media, environmental monitoring which means that growth in uncrewed aviation is not the growth of a sector. It is the acceleration of a horizontal capability that augments and transforms how dozens of industries do their most fundamental work.

From Hobby to Industry: The Regulatory Arc

Regulatory history is not, as a rule, gripping reading. But the story of how the United States built a legal framework for commercial drone operations is one of the more consequential policy narratives in modern American aviation: a story of contested authority, hard-won clarity, and a frontier that remains, in important respects, still open.

Before August 2016, commercial drone use existed in a legal gray zone that was uncomfortable for operators and bureaucratically awkward for regulators. The FAA had long asserted jurisdiction over the navigable airspace of the United States, but the rules it had developed over decades were designed for crewed aircraft: planes, helicopters, and the certified pilots who flew them. Small uncrewed systems occupied an uncertain legal space. Hobbyists flew under community-based safety guidelines that carried no statutory enforcement weight. Commercial operators, anyone who wanted to use a drone for compensation or hire, technically needed FAA authorization that was difficult to obtain and inconsistently enforced. Pioneering real estate photographers, agricultural consultants, and infrastructure inspectors operated at legal risk, in a market that was clearly developing faster than the rules governing it.

The FAA's Part 107 rulemaking, which took effect on August 29, 2016, changed that.[8] It was the most significant aviation rulemaking in decades for its category: a comprehensive legal framework for commercial UAS operations involving aircraft weighing under 55 pounds. Part 107 defined where drones could fly, how high, at what time of day, and under what operational constraints. It required commercial operators to obtain a Remote Pilot Certificate by passing a knowledge test covering airspace rules, weather interpretation, radio communications, and UAS regulations, a test that demanded real aviation literacy. You could not simply buy a drone and declare yourself a professional. You had to know the airspace you were entering.

That certification requirement was, in retrospect, both the industry's most important professionalization milestone and its first genuine talent pipeline challenge. Suddenly, a functioning commercial drone market required not just operators, but educated, tested, certified operators. The knowledge threshold was intentional and appropriate. It was also a structural bottleneck that the education system was not prepared to address at scale.

Rules clear the path. People and machines have to walk it.
And in the story of uncrewed aviation, it is the people question
that has never received the urgency it deserves.

The years following Part 107 brought a series of further regulatory developments that progressively built out the accountability and operational architecture of the industry. Remote ID, widely described as the "license plate for drones" was the next major milestone: a rule requiring drones to broadcast their identity, location, and operator location in real time, creating the infrastructure of airspace accountability. Remote ID compliance became mandatory on September 16, 2023, after years of rulemaking and industry debate.[9] Its significance goes beyond compliance: Remote ID is the data layer on which future traffic management systems depend, the prerequisite for the scaled, routine, mixed-use airspace operations that the industry's economic potential requires.

UAS Traffic Management (UTM) represents the next layer of that architecture. Developed through a multi-year NASA research program, UTM is the digital equivalent of air traffic control for the low-altitude drone world: a distributed, automated system for managing the flow of uncrewed aircraft through shared airspace without the continuous intervention of human controllers. UTM is not yet fully operational as a national system, but its frameworks have shaped how the FAA, industry, and technology providers think about a future sky populated by thousands of simultaneous drone operations.

The most consequential regulatory development in years arrived on August 7, 2025, when the FAA published its long-awaited Notice of Proposed Rulemaking (NPRM) for Beyond Visual Line of Sight (BVLOS) operations, a document exceeding 700 pages that proposed a new Part 108 framework designed to replace the existing waiver-based system for BVLOS missions with scalable, performance-based regulations.[5] BVLOS is the operational capability that unlocks the full economic potential of commercial drones: the ability to fly beyond the pilot's direct line of sight, across infrastructure corridors, over agricultural surveys spanning thousands of acres, and through delivery routes that extend far beyond any single operator's visual range. Without BVLOS authority, the economic model for drone delivery, long-haul inspection, and large-scale agricultural survey remains constrained. With it, the addressable market for uncrewed aviation expands by an order of magnitude.

The comment period on the BVLOS NPRM closed in October 2025, after the FAA received approximately 3,100 public comments, a figure reflecting both the breadth of industry interest and the depth of stakeholder concern about specific provisions.[5] Industry associations raised concerns about the treatment of existing Part 107 BVLOS waiver holders, the proposal's emphasis on high automation over pilot-in-the-loop control, and the regulatory burden its requirements would place on small operators and public safety agencies.[6] A final rule is expected to take years to implement. The frontier remains open.

What the regulatory arc of uncrewed aviation demonstrates, across a decade of hard-won progress, is that the United States has been serious about building the legal infrastructure for this industry, more serious, arguably, than any other major economy. But the arc also reveals a persistent pattern: rules clear the path. People and machines have to walk it. And in the story of uncrewed aviation, it is the people question that has never received the urgency it deserves.

Every Industry, Every Altitude

To understand why uncrewed aviation is different from most technology revolutions, it helps to spend time with the breadth of its application base, not as a list, but as a landscape. No single vertical dominates the story. The disruption is horizontal, cutting across industries with very different economics, risk profiles, and regulatory environments, but arriving in each with a similar promise: lower cost, higher frequency, better data, and less human exposure to dangerous conditions.

Agriculture was one of the first commercial sectors to adopt drone technology at meaningful scale, and it remains one of the largest application areas globally. Precision agriculture drones equipped with multispectral sensors can map crop health, soil moisture content, and pest pressure across thousands of acres in a single flight cycle, generating granular, field-level data that allows agronomists to prescribe inputs at the plant level rather than the field level. Drone spraying systems deliver fertilizers and pesticides with GPS-guided precision that reduces chemical use and minimizes runoff. This is not a future application being evaluated in research plots. It is current standard practice on forward-thinking farms across the Midwest, the Southeast, and the agricultural West.

Infrastructure inspection represents one of the strongest economic cases for drone deployment, precisely because the comparison set is so expensive and so dangerous: crewed inspection methods require significant coordination, cost, and risk. Power lines, natural gas pipelines, bridges, cell towers, wind turbines: all are being inspected by drones at lower cost, higher frequency, and with dramatically reduced risk to human workers. A licensed drone operator can inspect a cell tower in 20 minutes that would require a tower climber half a day and carry a meaningful probability of injury. Utility companies across the country are building internal UAS programs or contracting with specialized UAS service providers who deploy at scale. The thermal imaging, LiDAR (Light Detection and Ranging) point clouds, and high-resolution video that drone inspections produce are, in many cases, superior to what human inspectors could collect from the ground or from a rope.

Drone delivery has moved, more quietly than the headlines suggested it would, from pilot program to operational reality. As of April 2025, seven companies hold FAA Part 135 Air Carrier Certificates authorizing commercial drone delivery at scale: Wing (Alphabet's subsidiary), UPS Flight Forward, Amazon Prime Air, Zipline, Causey Aviation Unmanned, DroneUp, and DEXA (Drone Express)—a Dayton, Ohio-based company that also holds a nationwide BVLOS waiver. Each arrived at this regulatory threshold through years of development, FAA review, and operational demonstration. The BVLOS rulemaking, when finalized, is broadly understood to be the catalyst for the next phase of expansion: the moment when drone delivery stops being a premium curiosity in select markets and becomes a logistics infrastructure component that reshapes last-mile economics.

Public safety applications have emerged as one of the most compelling use cases, with some of the most concrete outcome data available. Drones as First Responder (DFR) programs, in which uncrewed aircraft are launched from fixed stations and arrive at emergency scenes before ground units, are now operating in dozens of American cities. The Chula Vista Police Department in California, one of the earliest and most studied DFR programs in the country, has demonstrated faster scene response times, improved situational awareness for responding officers, and better outcomes across a range of incident types. The program, launched in October 2018, surpassed 25,000 missions in 2026, with drones arriving on scene before ground units in more than 17,000 of those calls.[13] Fire departments use drones to map wildfire perimeters and identify hot spots. Search-and-rescue operations use thermal imaging drones to locate missing persons in terrain that ground teams cannot quickly reach.

The defense and national security applications of uncrewed aviation deserve their own book, and have received several. What is relevant here is the scale: CBP, the Coast Guard, and the Department of Defense (DoD) all operate substantial UAS fleets, with investments that are accelerating rather than stabilizing. DoD's interest in uncrewed systems spans the full spectrum: from small tactical reconnaissance platforms that individual soldiers can carry and deploy to large Group 5 systems capable of long-endurance intelligence, surveillance, and reconnaissance missions. The defense demand signal for domestic drone manufacturing is, as discussed later in this book, one of the most important economic forces shaping the emerging American UAS industrial base.

The architecture, engineering, and construction (AEC) industry has adopted drone surveying with a speed that surprised even its most optimistic advocates. Point cloud data generated from LiDAR equipped drones, orthomosaic maps produced from photogrammetric surveys, and 3D site models built from hundreds of overlapping drone images are replacing ground survey crews for many project types, compressing weeks of survey work into hours and producing deliverables that integrate directly into Building Information Modeling (BIM) software and design workflows. The AEC industry is not using drones because they are interesting. It is using drones because they are faster and cheaper: the only two arguments that have ever successfully displaced an established construction practice.

Even aerial cinematography, which might seem like a narrow creative application compared to infrastructure and agriculture, carries significant economic implications. Shots that required helicopter rigs costing thousands of dollars per hour are now available to any production with a licensed Part 107 operator and a prosumer drone. The democratization of aerial imagery has changed what small productions can afford to capture, what advertising campaigns can execute, and what news organizations can document. The full economic and creative implications of this shift are still unfolding.

Ready for Takeoff: But Who's Flying?

Every industry in the preceding section has something in common. Each is deploying a technology that is already capable enough to transform its operating model. The sensors are sharp, the batteries are improving, the autonomy stack is advancing, and the regulatory framework, while incomplete, is being built with more seriousness and speed than the early history of any comparable technology might have predicted. By the measures that technology analysts typically apply to assess whether an industry is ready to scale, uncrewed aviation passes comfortably. The technology is not the bottleneck.

What threatens to cap the industry's growth, and hand strategic advantage to foreign competitors who are moving fast and investing strategically, is a set of systemic gaps that are not technological in nature. They are human, financial, and industrial: the gaps that appear when a technology outpaces the organizational and institutional capacity required to deploy it at scale. There are three of them, and they are the subject of this book.

The first is the workforce gap. The uncrewed aviation industry cannot scale without people: not just remote pilots, though demand for certified, skilled, experienced pilots continues to grow faster than training pipelines are producing them, but the full spectrum of workers the industry requires. UAS technicians who can maintain and repair increasingly sophisticated platforms in the field; software engineers who build the autonomy and data processing systems that make drone operations useful; data scientists who turn raw drone-collected imagery into actionable business intelligence; fleet managers who coordinate multi-aircraft operations across enterprise deployments; regulatory affairs professionals who navigate an evolving and complex approval environment; and operations managers who understand both the technology and the business context in which it operates. Industry projections point toward hundreds of thousands of new workers needed in the coming decade. The education system is behind. The certification framework is young. And the talent pipeline is narrow in ways that are not merely inefficient but actively inequitable, failing to reach communities where capable workers exist but pathways into the industry do not.

The second is the investment misalignment gap. Billions of dollars are flowing into the drone ecosystem, as the funding round data cited earlier makes clear.[4] But capital follows hardware and software because hardware and software produce the intellectual property, the patents, and the revenue multiples that venture investors are built to chase. What it does not flow toward, with anything like proportionate urgency, is the workforce infrastructure, the domestic manufacturing capacity, and the regional ecosystem development that would make the industry's growth sustainable and strategically defensible. This is not a failure of the market, exactly. It is the market operating exactly as designed: optimizing for individual return rather than systemic resilience. Addressing it requires a different kind of intentionality: from industry leaders, from investors who hold a longer time horizon, and from a government sector that understands industrial strategy as something more than regulation.

More than 96 percent of drone operations in the U.S. national airspace are conducted on platforms manufactured by a single Chinese company. This is not merely a market share statistic. It is an economic dependency on a foreign-controlled supply chain

The third gap is the manufacturing vulnerability, and it is the one that has attracted the most urgent attention from policymakers in recent years.

For good reason. According to the FAA's ASSURE (Alliance for System Safety of UAS through Research Excellence) Center of Excellence 2025 Annual Report, published January 15, 2026, more than 96 percent of drone operations detected in the U.S. national airspace are conducted on platforms manufactured by DJI, the Chinese drone company whose dominance in the consumer and prosumer market has made its products the default choice across commercial applications as well.[7] Skydio, the leading American manufacturer, accounted for slightly more than one percent of detected platforms. Every other manufacturer combined represented less than 2.4 percent.[7] This is not merely a market share statistic. It is an economic dependency on a foreign-controlled supply chain, and a strategic vulnerability of the kind that the last decade of supply chain disruptions should have taught American industry to treat with seriousness.

The policy response has arrived: National Defense Authorization Act (NDAA) restrictions, the Blue UAS approved vendor framework, DoD-curated lists of domestically sourced platforms, and, as of December 2025, a Federal Communications Commission (FCC) ruling blocking new equipment authorizations for foreign-manufactured drones.[11][12] These actions have created a powerful demand signal for domestic drone manufacturing. But building an American drone manufacturing ecosystem requires workforce, capital, time, and coordination across the public and private sectors: exactly the kind of systemic, patient investment that markets alone do not generate.

These three gaps workforce, investment, and manufacturing are not independent problems that can be addressed sequentially or in isolation. They are a system. Solving the manufacturing gap requires the workforce to staff the factories. Building the workforce requires the investment to fund the training programs. Attracting the investment requires the market clarity that comes from a functioning domestic manufacturing sector producing platforms that buyers want to purchase. The feedback loops run in every direction. The only effective response is a systemic one.

A Defining Decade

The decisions being made right now in corporate boardrooms, in congressional budget negotiations, in university curriculum committees, in the offices of state workforce development agencies, and in the factories that are being planned or not planned in communities across the United States will determine whether America leads or follows in one of the defining technology sectors of the 21st century. This is the argument of this book. It is not a prediction about technology. The technology is already here. It is an argument about institutions, strategy, and the political will to build what the moment requires.

Underlying every application described in this book, and accelerating every economic projection attached to them, is artificial intelligence. Autonomous flight path planning, computer vision systems that identify infrastructure faults from thermal imagery, machine learning models that translate raw multispectral data into prescriptive agronomic recommendations, the supervisory architectures that will allow a single operator to manage hundreds of simultaneous aircraft under Part 108: none of this is speculative. It is operational or rapidly becoming so. This book treats AI as the engine: the force that makes drone economics transformational rather than merely useful, while focusing its argument on the institutional infrastructure required to deploy that engine at scale. The autonomy stack improves whether or not the workforce keeps pace. Whether or not the factories are built. Whether or not the capital flows to the right places. What does not happen automatically is the human, financial, and industrial architecture that turns AI-enabled drone capability into a functioning American industry. That gap, not the technology itself is what this book is about.

The author has had a front-row seat to both the extraordinary momentum of this industry and its persistent structural limitations. As CEO of Women and Drones since its founding, she has watched companies struggle to fill positions that did not exist five years ago. She has sat in investment meetings where human capital was mentioned as an afterthought after the discussion of battery chemistry and sensor specifications. She has seen the factories that have not been built yet because the demand signal was not clear enough, soon enough, or because the workforce to staff them does not exist in the communities where they would most naturally be located. She has seen the pilots who did not complete the certification process, not because they lacked the aptitude, but because they lacked access to the preparation that would have made the difference. The industry cannot build the talent pipeline it needs by fishing in the same narrow pool that every other technology sector is already competing over.

This book is not about drones. It is about what it takes to build an industry: the people, the capital, the policy environment, and the manufacturing base that turn a technology into an economy. Uncrewed aviation is the vehicle. Systemic industrial strategy is the destination. The drone sector offers the United States a generational opportunity to demonstrate that it can still build things, train people at scale, and compete on the technological frontiers that will define economic and national security for decades. The question this book attempts to answer is not whether that opportunity exists. It clearly does. The question is whether the country has the institutional capacity, the strategic patience, and the vision to seize it.

The next decade will answer that question. The answer is not yet written.

Looking Ahead — Chapter Two

Chapter Two, "Who Owns the Sky? The Geopolitics of Uncrewed Systems," turns to the global competitive landscape: the strategic investments that China has made in its domestic drone industry, the policy responses taking shape in Washington, and the uncomfortable reality that the United States is not building this industry in a vacuum. It is building it in competition. And the most important competitor is not a startup or an incumbent aerospace prime. It is a nation.

Notes and Sources

1. MarketsandMarkets. Unmanned Aerial Vehicle (UAV) Market — Global Forecast to 2030. 2025. Global UAV market valuation of $26.12 billion (2025), projected $40.56 billion (2030), CAGR 9.2 percent annually. Fully autonomous operations identified as the highest-growth segment.

2. Berg Insight / ResearchAndMarkets. Commercial Drone Market Report. 2025. Commercial drone installed base: 2.8 million units worldwide (2024), projected 4.5 million (2029).

3. Federal Aviation Administration. U.S. Civil Airmen Statistics, December 31, 2025. FAA Aeronautical Center, Oklahoma City, OK. Active Part 107 Remote

Pilot Certificates: 492,311. Total active pilots: 887,519. Student pilots: 370,286. Non-student traditional pilots (calculated): approximately 517,233.

4. StartUs Insights. Drone Industry Analysis & Global Startup Report. 2025. Global drone industry workforce: more than 2.1 million workers. Investment data: average $27.2 million per funding round; 2,000+ investors; 7,000+ funding rounds.

5. Federal Aviation Administration. Notice of Proposed Rulemaking: Normalizing Unmanned Aircraft Systems Beyond Visual Line of Sight Operations (Part 108). 90 FR 38212. Docket No. FAA-2025-1908. Published August 7, 2025. Comment period closed October 6, 2025. Approximately 3,100 public comments filed.

6. DLA Piper. "FAA's Proposed Part 108 BVLOS Rule: Industry Response and Key Concerns." October 13, 2025. Summary of stakeholder comments including concerns about transition from Part 107 waivers, automation requirements, and small operator burden.

7. FAA Center of Excellence for UAS Research — ASSURE (Alliance for System Safety of UAS through Research Excellence). ASSURE A83 2025 Annual Report. Contributing institutions: Embry-Riddle Aeronautical University, Kansas State University, NIAR, and Wichita State University. Published January 15, 2026. DJI platforms: 96%+ of detected operations in U.S. national airspace. Skydio: slightly more than 1%. All other manufacturers combined: less than 2.4%.

8. Federal Aviation Administration. Part 107 — Small Unmanned Aircraft Systems; Final Rule. 14 CFR Part 107. Effective August 29, 2016. Established the primary legal framework for commercial UAS operations in the United States.

9. Federal Aviation Administration. Remote Identification of Unmanned Aircraft; Final Rule. 86 FR 4390. Mandatory compliance date: September 16, 2023. Established real-time UAS identity and location broadcast requirements.

10. TCB Drones LLC / FAA Testing Data. "Part 107 Pass Rates, 2025." Published May 2025. Approximately 74,000 Part 107 knowledge test attempts in 2025; pass rate 82.96 percent.

11. U.S. Department of Defense / Department of Homeland Security. National Defense Authorization Acts (FY2020–FY2025). Provisions restricting procurement of covered foreign-manufactured UAS and establishing the Blue UAS Framework for approved domestic platforms.

12. Federal Communications Commission. Order Restricting New Equipment Authorizations for Covered Foreign-Manufactured Drone Platforms. Issued December 22, 2025.

13. Chula Vista Police Department. Drone as First Responder Program Milestone Data. As of May 2026: 25,000+ missions completed; drones arrived on scene before ground units in 17,170 calls; average first-on-scene response time 96.98 seconds. Sources: CVPD press release (May 2026); DroneXL (May 15, 2026); FOX 5 San Diego.

Who Owns the Sky? The Geopolitics of Uncrewed Systems

"Geopolitical competition is not only won with weapons and armies. It is won with supply chains, semiconductors, and the machines that carry data."

— Sharon Rossmark

The Day the Supply Chain Was Cut

In October 2024, the Chinese government did something remarkable, not because it was loud, but because it was almost entirely silent. No press conference. No formal proclamation. No trade war headline. Instead, acting under the authority of China's newly expanded export control and sanctions frameworks, government officials imposed direct sanctions on Skydio, the Redwood City, California-based manufacturer that had become the United States' leading domestic drone producer.[1]

The target was specific and strategically chosen: batteries. Chinese authorities ordered Dongguan Poweramp, a TDK subsidiary and Skydio's primary battery supplier, to immediately cease all business with the company. Not airframes, not software, not motors: batteries, the one critical component that Skydio had not yet fully sourced from domestic or allied manufacturers. The intent was unmistakable. Cripple the supply of the one input that no drone can fly without, and you ground America's most visible domestic drone manufacturer overnight.

It did not fully achieve its aim. Skydio had contingency plans. The company moved rapidly, secured alternative supply relationships, and emerged from the episode having rebuilt its supply chain outside of China. As CEO Adam Bry wrote to customers at the time: "This is a clarifying moment for the drone industry. If there was ever any doubt, this action makes clear that the Chinese government will use supply chains as a weapon to advance their interests over ours."[1]

But the episode revealed something that policymakers, investors, and industry leaders could no longer ignore. The drone industry is not merely a commercial technology sector. It is a geopolitical battleground. And for the United States, the terrain was far more exposed than anyone had publicly acknowledged. The sanctions had identified the exact pressure point in an American company's supply chain with a precision that reflected how thoroughly China understood American industrial dependencies, knowledge accumulated in part because American companies had spent years building their supply chains inside China's borders.

Understanding who owns the sky, then, does not begin with looking upward at the airspace. It begins with looking downward: into the supply chains, the data architectures, the component dependencies, and the manufacturing relationships that determine who actually controls the machines filling it. The story of drone geopolitics is, at its core, a story about what happens when a nation discovers, later than it should have, that it built its future on someone else's foundation.

Shenzhen's Quiet Conquest

To understand how the United States arrived at this moment, it is necessary to understand how DJI, Da-Jiang Innovations (meaning roughly "Great Frontier Innovations" in Mandarin), came to dominate the global drone market. That story is not, as it is sometimes told in Washington, a simple tale of Chinese industrial policy and unfair advantage. It is also a story of genuine technological achievement, brilliant market timing, and a set of regulatory conditions that the United States itself created.

DJI was founded in 2006 by Frank Wang, known in Chinese as Wang Tao, while studying at the Hong Kong University of Science and Technology (HKUST). Wang was an aviation enthusiast who had grown frustrated with the instability of remote-controlled helicopters, and his early work, developed in his university dormitory, focused on the flight control systems that could make small aircraft hover reliably. He moved his operation to Shenzhen, the Pearl River Delta city that had already produced Huawei, ZTE, and one of the world's most vertically integrated electronics manufacturing ecosystems. The geography mattered enormously. Shenzhen's manufacturing supply chains, the density of component makers, the speed of prototyping, the cost of production, gave DJI capabilities that no Western startup could easily replicate.[2]

In its early years, DJI sold flight stabilization systems to hobbyists and advanced model aircraft builders. They were genuinely superior products. When DJI introduced the Phantom consumer drone in 2013: a ready-to-fly quadcopter with integrated camera gimbal stabilization that could be purchased, unboxed, and airborne within minutes. It fundamentally democratized aerial photography. The Phantom was not the first consumer drone; it was simply the first one that worked well enough, consistently enough, at a price point low enough, that ordinary people could use it without specialized training.

The commercial inflection arrived in August 2016, when the Federal Aviation Administration (FAA) implemented Part 107 of the Federal Aviation Regulations, establishing the first comprehensive regulatory framework for commercial small drone operations in the United States. Part 107 opened a vast new market. Photographers, filmmakers, construction companies, insurance adjusters, agricultural operators, real estate agencies, and industrial inspection firms suddenly had a legal pathway to put drones to work. When they went looking for hardware, DJI was waiting. The company had spent years building not just aircraft but an ecosystem: software, flight planning tools, accessories, training resources, and a global service network. Enterprises did not merely buy DJI drones; they built their entire operational workflows around them.

America built the runway. China's drone industry took off on it.

The market dominance that resulted is, by any measure, extraordinary. DJI today holds approximately 80 percent of the United States commercial drone market and more than 90 percent of the global consumer drone market.[3] In 2025, DJI platforms accounted for 83.48 percent of all drone detections globally, a figure that includes detections across conflict zones and operational military theaters, not merely civilian commercial environments.[4] These are not the statistics of a company that competes in a market. They are the statistics of a company that, for most practical purposes, is the market.

The "stickiness" of DJI's position is a dimension of this dominance that is frequently underestimated in policy discussions. Switching away from DJI is not analogous to choosing a different brand of laptop or smartphone. Operators trained on DJI platforms, certified on DJI software, and integrated into DJI's data ecosystems face a genuine and costly transition. Certifications must be renewed on new systems. Operational procedures must be rewritten. Data workflows must be rebuilt. The institutional knowledge embedded in years of DJI-based operations is not easily transferred. DJI has understood this dynamic, investing heavily in the software and services layers that deepen customer dependency far beyond the hardware itself.

The economic entanglement is equally significant. DJI's products and services contribute an estimated $116 billion in economic value to the United States economy and support nearly 450,000 American jobs, a figure derived from DJI-commissioned analysis deployed in the company's lobbying efforts against procurement restrictions.[5] That figure carries the limitations of its source and should be read with appropriate skepticism about methodology, but the underlying reality it reflects is not fabricated: hundreds of thousands of American businesses, operators, and workers have built livelihoods around Chinese-manufactured hardware. From 2010 to 2022, on average 85 percent of drones purchased by United States state agencies were Chinese-made, a dependency constructed through years of procurement decisions that prioritized price and performance over the question of where the hardware was actually made and who controlled its software.[6]

The irony of this situation is sharp and instructive. The United States government, through Part 107, created the open commercial market conditions that enabled DJI's dominance. American consumers, companies, and government agencies, exercising entirely rational purchasing decisions based on price and capability, funded DJI's growth into a multi-billion-dollar enterprise. America built the runway. China's drone industry took off on it.

From Market Share to National Security: Two Arguments That Must Not Be Confused

The United States government's reckoning with Chinese drone dominance did not arrive all at once. It came in stages, each more consequential than the last, across a period that stretched from an early institutional signal widely dismissed as protectionism to a comprehensive legislative and regulatory architecture that now reaches into every corner of federal procurement. But to understand both the strength and the limits of the policy response that architecture represents, it is essential to distinguish between two fundamentally different national security arguments that have too often been conflated in the Washington debate.

The first is the active threat argument: the claim that DJI platforms contain hidden backdoors, transmit operational data to Chinese government servers, or can be weaponized or hijacked by foreign actors. This argument has driven much of the political rhetoric surrounding DJI restrictions and has been repeated in congressional testimony, Department of Homeland Security (DHS) advisories, and public advocacy by domestic drone manufacturers. It is also the argument that has attracted the most serious evidentiary scrutiny. And the most significant challenge.

In May 2026, U.S.-based cybersecurity firm OnDefend published the results of the most comprehensive independent technical assessment ever conducted on DJI drone systems. Over a five-month period, OnDefend tested the DJI Air 3S and DJI Matrice 4E platforms along with their controllers, firmware, applications, RF systems, and communications protocols. The methodology was rigorous: software and firmware analysis, hardware teardown, RF spectrum analysis covering 1 MHz to 6 GHz, network traffic monitoring, supply chain verification, adversarial attack simulations, replay and jamming testing, and jailbreak attempts. The firm's clients include government entities, lending the assessment credibility that no DJI-commissioned study could carry.[14]

The findings were direct: no evidence of hidden backdoors, no unauthorized transmission of data outside the United States, and no viable pathways for hijacking or weaponizing the tested platforms. DJI's Local Data Mode was found to successfully prevent user flight data from being transmitted to internet-based destinations while enabled. No evidence was found that historical flight data was later uploaded once systems returned to normal operating mode, directly contradicting claims that had been made in congressional testimony. Only ten low-risk findings and thirteen observations emerged from the assessment, none representing realistic threats to flight safety or widespread data exposure.

The OnDefend assessment does not end the policy debate. Cybersecurity is never finished; firmware changes, software evolves, and threats shift. The assessment evaluated specific platforms at a specific point in time and recommended continuous validation rather than one-time certification. But it does establish that when subjected to extensive independent technical testing, the catastrophic active-threat scenarios that have dominated the Washington conversation were not supported by the evidence. Any honest treatment of the national security argument must acknowledge that.

The second argument, the structural vulnerability argument, is distinct from the active threat framing and is not weakened by the OnDefend findings in any way. This argument does not require DJI to be acting in bad faith today. It requires only that the United States recognize what the Skydio battery episode demonstrated without any ambiguity: China does not need backdoors to disrupt American drone operations. It can simply turn off the supply chain. The structural vulnerability argument is about dependency: what it means for a nation to have built 96 percent of its drone operational capacity on platforms manufactured by a single foreign company, sourced from component supply chains that the same foreign government controls at the raw material level. That dependency is a strategic liability regardless of DJI's current behavior, because the liability is not about what DJI is doing. It is about what China can do.

China does not need backdoors to disrupt American drone operations. It demonstrated that clearly when it turned off Skydio's battery supply. The structural vulnerability argument has never required DJI to be acting in bad faith. It requires only that the United States recognize what dependency actually means.

These two arguments demand different policy responses. The active threat argument, if it were supported, would justify immediate operational restrictions on all DJI platforms across all contexts. The structural vulnerability argument justifies a deliberate, phased transition away from foreign-controlled manufacturing dependency, one that takes seriously the disruption costs to the commercial operators and public safety agencies who built their programs on DJI platforms, and one that invests in the domestic manufacturing capacity and component supply chains that genuine sovereignty actually requires. Conflating the two arguments has produced a policy response calibrated to an active threat that independent technical assessment has not confirmed, applied to a commercial operator community that bears the cost of restrictions without having received a credible domestic alternative.

The first significant institutional warning came in August 2017, when the United States Army suspended its use of DJI drones, citing operational security and safety concerns. The decision was greeted skeptically in some quarters, concerns the OnDefend assessment has now given more evidentiary weight than they had at the time. But the structural risk argument, about what it means to operate critical data collection infrastructure dependent on foreign-controlled supply chains, was always the more durable of the two, and it remains fully intact.

Drones are data collection platforms of extraordinary capability. A commercial drone conducting an infrastructure inspection simultaneously records GPS coordinates, high-resolution imagery, LiDAR (Light Detection and Ranging) or thermal sensor data, and complete flight telemetry. A fleet of drones operating in an urban environment over the course of months accumulates a detailed, precise, and continuously updated map of that environment: its infrastructure, its patterns of activity, its vulnerabilities. The question of where that data goes, who can access it, and what legal framework governs it is not hypothetical. It is a legitimate policy concern. What the evidence now suggests is that the answer to that question, for DJI platforms under properly configured operating conditions, is more reassuring than the most alarming claims in the Washington debate implied.

The legislative response built gradually through successive National Defense Authorization Acts (NDAAs) before reaching its decisive moment. Section 848 of the FY2020 NDAA prohibited the Department of Defense (DoD) from procuring certain categories of foreign-manufactured components. The American Security Drone Act, enacted on December 22, 2023 as part of the FY2024 NDAA and codified as Public Law 118-31, represented the most sweeping action to date: a prohibition on the use of any federal funds to procure or operate drones manufactured or assembled by covered foreign entities, effective two years from enactment.[7] The Federal Acquisition Regulation (FAR) interim rule that took effect on November 12, 2024 made this prohibition binding across all federal contracts awarded on or after that date.[8] And on November 21, 2025, the

Office of Management and Budget (OMB) issued a government-wide implementation memorandum establishing comprehensive secure UAS procurement requirements not only for federal agencies but for all recipients of federal funds, a provision that reaches state and local governments, universities, and nonprofits that operate on federal grants.[9]

The most consequential single action, in terms of immediate market impact, came in December 2025, when the FCC added DJI to its national security Covered List. The mechanism was striking in its procedural logic: the FY2025 NDAA required a formal national security review of DJI by relevant federal agencies. No agency completed that review before the statutory deadline. Under the terms of the legislation, the failure to complete the review triggered automatic listing, meaning DJI was added to the Covered List not because a review found it posed a threat, but because no review was conducted at all. The consequence was immediate and severe: new DJI products and imports are now blocked from receiving FCC radio frequency authorization, effectively preventing DJI from bringing new hardware to the United States market. DJI filed suit against the FCC in February 2026; the litigation is ongoing. Twenty-five products the company had planned to launch in the U.S. market in 2026 are frozen, representing approximately $1.5 billion in projected revenue.[10]

The cascade effects on state and local government are substantial and still unfolding. More than 1,200 public safety agencies, including police departments, fire departments, and emergency management offices, had built their drone programs on DJI platforms. They trained their operators on DJI systems. They integrated DJI software into their dispatch and incident management workflows. Federal restrictions are now beginning to flow into state and local procurement decisions, creating operational gaps that cannot be filled overnight and generating pressure on jurisdictions that had never thought of their drone programs as a national security matter at all. Compounding this pressure is a persistent confusion across agencies, vendors, and integrators about which systems are truly compliant with the applicable requirements, a confusion made worse by inconsistent terminology and by the active presence in the market of platforms marketed as domestic alternatives that still rely heavily on foreign-sourced components.

Table 2.1 — Key U.S. Legislative, Regulatory, and Technical Developments on Chinese-Manufactured UAS, 2017–2026

Date	Action	Significance
August 2017	U.S. Army suspends DJI drone use	First major institutional security signal; widely debated at the time
December 22, 2023	American Security Drone Act enacted (NDAA FY2024, P.L. 118-31)	Prohibits federal funds for covered foreign-entity drone procurement
November 12, 2024	FAR interim rule takes effect	DoD, GSA, and NASA barred from covered UAS in any new contract
November 21, 2025	OMB government-wide implementation memorandum	Requirements extended to all federal agencies and grant recipients
December 2025	FCC adds DJI to national security Covered List	New DJI products blocked from FCC radio frequency authorization; triggered by failure to complete statutory review, not by review findings
February 2026	DJI sues FCC; litigation ongoing	25 products and approximately $1.5 billion in planned U.S. revenue frozen
May 2026	OnDefend independent security assessment published	No evidence of hidden backdoors, unauthorized data transmission, or weaponization pathways found in DJI Air 3S and Matrice 4E platforms

Sources: Public Law 118-31; FAR interim rule; OMB Memorandum, November 21, 2025; FCC Covered List, December 2025; OnDefend DJI Security Assessment, May 2026.

The Ground Beneath the Ground

The legislative arc traced in the previous section represents a genuine and necessary policy response to a real national security exposure: specifically, the structural vulnerability exposure. But it contains an assumption that, if left unexamined, will cause the United States to mistake the symptom for the disease. The assumption is this: that the problem is primarily one of finished products: that if the United States stops buying Chinese-assembled drones, it has solved the sovereignty problem. It has not. Not even close.

The supply chain problem runs far deeper than finished goods. China controls approximately 90 percent of global rare earth mineral processing, the materials essential to the sensors, motors, and electronic components found in virtually every drone regardless of where the final assembly occurs.[12] Rare earth elements such as neodymium, dysprosium, and terbium are not substitutes; they are the physical basis of the technologies that make modern drones possible. China's dominance in their processing is not a recent development. It is the product of decades of deliberate industrial policy, investment in processing infrastructure, and, in some cases, environmental practices that Western democracies have declined to replicate. That accumulated advantage cannot be dissolved by an act of Congress.

A drone with a U.S.-made airframe, Chinese sensors, Chinese batteries, Chinese motors, and Chinese software is not a domestically sovereign drone. It is a Chinese drone with American sheet metal.

The battery situation is even more stark. China produces approximately 99 percent of drone battery cells globally, not a figure that reflects competitive supply, but a near-total monopoly on a critical input.[12] When Chinese authorities targeted Skydio's battery supply in October 2024, they were not selecting arbitrarily from among several available pressure points. They were selecting the one chokepoint that they knew, with high confidence, that no American company had yet successfully routed around. Batteries were the kill switch.

The permanent magnet situation follows the same pattern. Approximately 90 percent of the permanent magnets that power drone motors are produced in China. These magnets, the neodymium-iron-boron alloys that make small, powerful electric motors possible, are not easily substituted. The manufacturing processes required to produce them at scale, the raw material access, and the production efficiencies that China has built over decades cannot be replicated quickly regardless of how strong the policy signal is. American manufacturers who want to build "American" drone motors face the reality that the magnets inside those motors almost certainly originated in China.[12]

This creates what might be called the "assembled in America" problem, and it is not merely theoretical. In the current market, responding directly to the procurement signal that the legislative restrictions have created, platforms are being marketed as domestic alternatives that still rely heavily on foreign-sourced components. The legislative framework as currently constructed does not fully address this behavior. The American Security Drone Act focuses primarily on the nationality of the manufacturer of the final assembled product, a necessary condition for addressing the most visible layer of the dependency, but an incomplete solution to the deeper supply chain vulnerability that the Skydio episode exposed. Agencies seeking to make compliant procurement decisions face persistent confusion about which systems genuinely reduce their foreign component exposure and which simply relocate their final assembly. A drone with an American-designed airframe, Chinese sensors, Chinese batteries, Chinese motors, and Chinese software is not a domestically sovereign drone. It is a Chinese drone with American sheet metal. The market currently offers limited tools for distinguishing one from the other with confidence.

The most serious effort yet to address this deeper problem is Skydio's SkyForge initiative, announced in April 2026 as part of a $3.5 billion, five-year investment commitment to expand U.S. drone manufacturing.[11] SkyForge is not simply a facility expansion announcement; it is a deliberate attempt to reconstruct the American drone supply chain from the component level upward. The program invites domestic suppliers to co-locate with Skydio's expanded manufacturing operations, directing more than $1 billion specifically to domestic component suppliers. The logic is straightforward: if the supply chain vulnerability runs multiple levels deep, the solution must also run multiple levels deep. You cannot solve a component problem by assembling components faster.

What the Skydio episode and the SkyForge initiative together reveal is a principle that will define the next decade of American drone policy: building a genuinely sovereign domestic drone manufacturing industry requires not just moving final assembly operations to American soil, but rebuilding the component supply chains — batteries, magnets, sensors, processors — from the ground up. That is a multi-year, multi-billion-dollar undertaking. It requires sustained investment, patient capital, and workforce capabilities that do not yet exist at the scale the industry needs. Skydio has begun it. The broader industry, and the government programs designed to support it, must follow.

Building the Alternative: Blue UAS, Green UAS, and the Framework Gap

Restricting Chinese-manufactured drones from federal procurement was the first move in the American government's drone strategy. But a prohibition without an alternative is a policy with a gaping hole. If the agencies, contractors, and first responders who depended on DJI platforms have nowhere to turn, the restrictions simply create operational dysfunction without advancing genuine security. The affirmative side of the U.S. government's drone strategy, the effort to build a credible domestic alternative, has produced two distinct procurement frameworks that serve different segments of the market and are frequently and consequentially confused with each other.

The first is NDAA compliance. Under Section 848 of the FY2020 NDAA and subsequent legislation, a drone is NDAA-compliant if it uses secure, approved hardware and avoids prohibited foreign-manufactured components. NDAA compliance is a legal threshold, not a list. It defines what the law requires; it does not independently certify that a specific platform meets those requirements.

The second is the Blue UAS Cleared List, administered by the Defense Innovation Unit (DIU). Blue UAS is the federal government's primary mechanism for signaling which manufacturers meet national security standards for DoD drone procurement. The process involves cybersecurity evaluation, NDAA compliance verification, and administrative documentation review. Manufacturers whose platforms pass are listed as cleared. All Blue UAS drones are NDAA-compliant, but not all NDAA-compliant drones are Blue UAS. Blue UAS is effectively a pre-approved-for-federal-use pathway: a DoD-specific certification that carries significant credibility for military and federal procurement officers. As of the August 2025 update, the Blue UAS list includes 23 drone platforms and 14 UAS framework components spanning a range of operational capabilities.[10]

This distinction matters because Blue UAS is designed for and calibrated to the federal and defense procurement environment. Its manufacturers are primarily pursuing defense contracts. Their platforms are priced and specified for government procurement. The working commercial operator, the agricultural consultant, the infrastructure inspector, the real estate photographer, the small public safety agency managing its first drone program, exists largely outside the market that Blue UAS was designed to serve.

This gap is where the third framework becomes critical: Green UAS. Developed by the Association for Uncrewed Vehicle Systems International (AUVSI), the Green UAS program provides a cybersecurity and supply-chain vetting pathway specifically designed for state, local, commercial, and critical infrastructure operators who need a trusted platform but do not require or qualify for the full rigor of DoD-level Blue UAS certification.[15] Green UAS bridges the gap between NDAA compliance as a legal floor and Blue UAS as a defense-specific ceiling, offering a scalable, industry-driven alternative for the non-federal users who represent the majority of the American drone operator community. For the more than 1,200 public safety agencies that built their programs on DJI and are now navigating the transition to compliant alternatives, Green UAS is the more directly relevant framework, and the one least discussed in the policy conversation that shaped the restrictions they are now navigating.

The manufacturers on the Blue UAS list represent the contours of the emerging defense-focused domestic alternative. Skydio appears with its X10D and X2D platforms. Parrot, the French manufacturer, is listed with its ANAFI UKR and ANAFI USA platforms, a reminder that "domestic alternative" in the drone context includes trusted allied-nation manufacturers, not only American companies. Inspired Flight, Freefly Systems, ModalAI, Wingtra, AeroVironment, Teal, Teledyne FLIR, Shield AI, and others complete the list, a roster that would have been unrecognizable to most drone procurement officers just five years ago, when DJI was the default answer to virtually every operational requirement.

The scale of global interest in this framework is telling. The DIU's 2025 Blue UAS Refresh Challenge drew 369 proposals from companies across 19 nations, a number that measures both the size of the opportunity the restrictions have created and the breadth of the global defense industrial base that has recognized the United States government's procurement signal. Companies in allied and partner nations understand that access to

American defense and federal markets requires meeting American security standards; the Refresh Challenge response demonstrates that the market is responding to that understanding.[10]

The policy environment for domestic drone manufacturing shifted significantly in the first half of 2025. On June 6, President Trump signed Executive Order 14307, "Unleashing American Drone Domination," directing federal support for domestic drone manufacturing and commercial UAS integration across the economy. Following the executive order, on July 10, 2025, Secretary of Defense Pete Hegseth signed a memorandum titled "Unleashing U.S. Military Drone Dominance," rescinding Biden-era policy restrictions that had slowed domestic drone procurement and development and delegating procurement authority directly to combat units.[13] Together, the executive order and the Hegseth memorandum represented the most explicit signal yet that the federal government intends to use the full weight of its procurement power to build the domestic industry.

The private sector has responded with the most significant investment commitment in American drone manufacturing history. Skydio's April 2026 announcement of a $3.5 billion, five-year investment plan, including a new manufacturing facility five times the size of its current space, more than 2,000 direct jobs, more than 3,000 supply chain jobs through the SkyForge program, and more than $1 billion directed to domestic suppliers, was accompanied by a Series F funding round of $110 million at a $4.4 billion company valuation.[11] Skydio has already delivered more than 60,000 drones to more than 3,800 customers, including more than 1,200 public safety agencies, every branch of the United States military, and 29 allied nations. That customer base and operational track record give the company a credibility that the investment community has clearly recognized.

Yet an honest assessment of where the Blue UAS ecosystem stands relative to what it must eventually replace requires acknowledging the gap that still exists. The DJI platform ecosystem, at its peak, offered dozens of purpose-built commercial products at a range of price points, supported by one of the most mature software stacks in the industry and a global distribution and service network. The Blue UAS list offers 23 platforms across a much narrower capability and price range, oriented almost entirely toward defense and enterprise procurement. The software maturity, operational variety, price accessibility, and global availability of the domestic alternatives lag significantly behind what DJI provided. The restrictions have created the demand signal. The supply response is underway. But the gap is real, and it falls most heavily on the segment of the market least equipped to absorb it.

Table 2.2 — Selected Manufacturers on the Blue UAS Cleared List (DIU, Updated August 2025)

Manufacturer	**Platform(s)**	**Country**	**Primary Use Case**
Skydio	X10D, X2D	United States	Defense, public safety, infrastructure
Parrot	ANAFI UKR, ANAFI USA	France	Defense, ISR, commercial operations
AeroVironment	Multiple platforms	United States	Defense, reconnaissance, logistics
Teledyne FLIR	Multiple platforms	United States	Defense, thermal imaging, public safety
Shield AI	Multiple platforms	United States	Defense, autonomous operations
Inspired Flight	IF800 series	United States	Defense, inspection, payload operations
Freefly Systems	Alta X	United States	Heavy-lift, cinematography, defense
ModalAI	VOXL platform	United States	Autonomy stack, GPS-denied navigation
Teal Drones	Golden Eagle	United States	Defense, ISR, first responder
Wingtra	WingtraOne GEN II	Switzerland	Mapping, surveying, precision agriculture

Full list includes 23 platforms and 14 UAS framework components across allied-nation and domestic manufacturers. Source: Defense Innovation Unit, Blue UAS Cleared List, August 2025.

The Consumer and Prosumer Gap: The Market No One Has Filled

There is a dimension of the domestic manufacturing challenge that the Blue UAS framework, the SkyForge announcement, and the policy conversation in Washington have almost entirely overlooked. It is the dimension that matters most to the majority of the 492,000-plus Part 107 commercial pilots whose operational livelihoods were built on platforms that are now restricted or unavailable and it is the dimension that reveals the deepest structural gap between the policy response and the market reality it was designed to address.

As of 2026, the United States does not produce a true mass-market consumer or prosumer drone. The lightweight, foldable quadcopters used for travel, recreation, real estate photography, small-scale agricultural monitoring, and the full range of commercial applications that Part 107 unlocked are manufactured almost entirely overseas, where vertically integrated supply chains for motors, electronic speed controllers (ESCs), gimbals, camera modules, and battery cells enable price points that U.S. manufacturers cannot currently match. American companies have concentrated their manufacturing efforts in the professional and enterprise tiers, where security, reliability, and mission-specific performance command premium prices and justify defense procurement budgets. This has left a wide unmet commercial demand in the middle market: the working drone operator who needs a capable, affordable, compliant platform and has found that none of the domestic alternatives answer that need at a price point their business model can support.

The implications extend well beyond individual operators. The consumer and prosumer gap affects STEM education, where early drone training relies almost entirely on foreign-made platforms. It affects the K-12 and community college workforce pipeline, where future pilots and technicians learn the craft on systems that are not domestically manufactured. It affects the innovation pathway that historically runs from consumer market competition through prosumer refinement into enterprise capability: a progression that cannot function if the consumer segment has no domestic participants. And it affects the small businesses and sole proprietors who make up a substantial share of the 492,000-plus certified remote pilot workforce, who built their operational workflows around DJI and have not been offered a credible domestic transition path at a comparable price point.

Several startups are exploring sub-$5,000 NDAA-compliant platforms, and there is early evidence of DoD interest in small domestically manufactured quadcopters that could seed the commercial market through defense procurement volumes. But as of the time of this writing, none have reached mass production. The consumer and prosumer segment remains the clearest evidence that the U.S. uncrewed aircraft ecosystem, while rapidly advancing in the defense and enterprise tiers, has not yet built the full-spectrum domestic capability that a genuine response to Chinese market dominance would require.

This gap is not merely a commercial inconvenience. It is a strategic one. The Part 107 commercial pilot workforce is the largest, most distributed, and most economically diverse segment of the American drone operator community. If the domestic manufacturing response serves only the defense contractor and large enterprise markets, it has solved the federal procurement problem while leaving the broader commercial ecosystem the operators who represent the industry's largest workforce and most direct public interface without a domestic alternative. That is a policy outcome that deserves to be named clearly, not papered over with procurement tables that count Blue UAS platforms without acknowledging what price tier and market segment those platforms actually serve.

The Long-Term Structural Challenge

The consumer and prosumer gap is the most visible manifestation of a set of broader structural challenges that will shape the long-term competitiveness and resilience of the U.S. domestic drone manufacturing ecosystem. These challenges are distinct from the supply chain dependency arguments made earlier in this chapter and deserve their own explicit framing.

The first is the absence of standardized testing and certification frameworks for U.S.-manufactured drones. Without consistent, publicly verifiable standards against which domestic platforms can be evaluated and compared, operators lack the tools to make confident procurement decisions and the confusion between NDAA compliance, Blue UAS certification, and Green UAS vetting that currently characterizes the market will continue to create conditions for inadvertent non-compliant procurement. Standardization is the infrastructure on which market confidence is built. The United States has not yet built it at the scale the domestic manufacturing ambition requires.

The second is investor hesitancy toward domestic drone startups, particularly in the consumer and prosumer segment where the capital requirements are substantial and the competitive disadvantage against Chinese manufacturers is most acute. The defense and enterprise tiers have attracted significant investment, including Skydio's $4.4 billion valuation and the SkyForge commitment. The commercial market tier, where the volume opportunity is largest and the strategic stakes for the broader operator community are highest, has attracted far less. Capital follows defense contracts because defense contracts provide revenue visibility. The commercial market, for all its scale, does not offer the same visibility and until it does, the investment gap will persist.

The third is fragmented state-level procurement rules, which create inconsistent pathways for agencies seeking to transition away from foreign-manufactured platforms. A public safety agency navigating a transition from DJI to compliant alternatives faces different requirements, timelines, and compliance definitions depending on which state it operates in and which federal funding streams it draws on. That fragmentation imposes compliance costs on precisely the agencies under-resourced municipal and county governments least equipped to absorb them.

The fourth is the workforce constraint, which connects directly to the argument of Chapter Three: limited training pipelines mean limited capacity to staff the manufacturing facilities, operate the platforms, maintain the fleets, and develop the software ecosystems that a genuine domestic manufacturing industry requires. The manufacturing ambition cannot outpace the human capital development that sustains it.

The fifth is the absence of a unified national strategy for scaling U.S. drone production. Executive Order 14307 and the Hegseth memorandum provided important directional signals. What they did not provide was the coordinated, cross-agency, public-private roadmap that translating those signals into manufacturing capacity at scale actually requires. The Creating Helpful Incentives to Produce Semiconductors (CHIPS) and Science Act provide an imperfect but instructive analogy: the semiconductor manufacturing investment enabled was accompanied by a degree of strategic coordination, research investment, and workforce development planning that the drone sector's executive-order-level direction has not yet matched.

These five structural challenges are not sequential problems to be addressed one at a time. They are a system. Standardization builds investor confidence. Investment funds workforce development. Workforce development enables manufacturing scale. Manufacturing scale generates the market evidence that allows state procurement frameworks to cohere around common standards. The feedback loops run in every direction — and the only effective response, as this book argues from its first pages, is a systemic one.

Why This Is Bigger Than Drones

The contest over drone manufacturing sovereignty is not an isolated industrial policy debate. It is a microcosm of the broader technology competition between the United States and China, one that illuminates both the strategic logic driving American policy and the depth of the structural challenge that policy must overcome.

The closest historical analogy is semiconductors. Beginning in the late 2010s, the United States moved systematically to restrict Chinese access to advanced semiconductor technology: export controls on chip design software, restrictions on advanced chipmaking equipment, prohibitions on the sale of cutting-edge processors to Chinese entities. The parallel domestic investment followed: the CHIPS and Science Act of 2022 committed more than $52 billion to rebuilding American semiconductor manufacturing capacity. But in semiconductors, the United States was the incumbent dominant power: home to the American supply chain partners of ASML, the Dutch chipmaking equipment company whose lithography machines are essential to advanced chip production, as well as Applied Materials, Lam Research, and a generation of fabrication expertise that China is still struggling to replicate. In drones, the dependency runs in precisely the opposite direction. China makes the components. America buys them. The United States is not the incumbent restricting a challenger;

it is the dependent party attempting to reconstruct its own industrial base from a position of structural weakness.

The lessons of Ukraine have made this problem impossible to ignore at the strategic level. The war that began in 2022 demonstrated definitively that commercial drones are battlefield assets, not merely civilian tools. DJI platforms were used extensively by both Ukrainian and Russian forces in the conflict's early phases before operational and political pressures led to restrictions. The proliferation of first-person-view (FPV) racing drones, built almost entirely from Chinese-manufactured components and adapted for strike and reconnaissance missions at a cost of a few hundred dollars per unit, reshaped tactical warfare in ways that no military planner had fully anticipated. The military and geopolitical utility of cheap, mass-producible uncrewed systems is no longer a theoretical projection. It has been demonstrated at scale, in real combat, against real military formations. Any nation that cannot produce those systems domestically at scale, at cost, with supply chain independence has accepted a strategic liability.

The allied dimension of this challenge is equally significant and less frequently discussed. Australia, the United Kingdom, and NATO member states are grappling with questions about drone sovereignty that are structurally identical to those facing the United States. As the Refresh Challenge participation described earlier in this chapter demonstrates, the allied security community has broadly recognized the demand signal that American policy has created. International coordination on drone security standards is beginning to emerge, but it remains nascent and uneven. The United States has an opportunity to lead the development of allied-nation standards for trusted UAS procurement—an opportunity that would simultaneously strengthen the domestic industrial base and deepen the security cooperation that allied drone sovereignty requires.

The workforce dimension, however, is where the geopolitical argument meets its most practical constraint and where the strategic discussion has been most notably incomplete. Every manufacturing facility that Skydio is planning, every domestic supplier that the SkyForge program intends to cultivate, every Blue UAS platform that a federal agency might choose over a DJI alternative, every consumer-tier platform that the market still needs someone to build, requires human beings to design it, build it, test it, certify it, operate it, and maintain it. The machines do not build themselves. The supply chains do not reorganize themselves. The software ecosystems do not write themselves. The geopolitical case for drone sovereignty is compelling, well-documented, and increasingly well-funded. What it has not yet adequately addressed is the human capital question: the question of whether the United States has, or is building, the workforce that manufacturing sovereignty across the full spectrum of the market actually demands.

Looking Ahead — Chapter Three

The geopolitical argument of this chapter points toward an unavoidable conclusion: building a domestically sovereign drone industry is not primarily a technology problem or a policy problem. It is a people problem. Chapter Three asks the question that the geopolitical debate has almost entirely overlooked: who are the people who will build this future, and are they being trained?

Notes and Sources

1. Multiple verified sources: TechCrunch; DroneDJ; DroneXL; Asia Times; DRONELIFE. October–November 2024. Chinese sanctions imposed October 11, 2024. Dongguan Poweramp (TDK subsidiary) ordered to cease all business with Skydio. Verified Adam Bry direct quotes from October 2024 customer letter: "This is a clarifying moment for the drone industry. If there was ever any doubt, this action makes clear that the Chinese government will use supply chains as a weapon to advance their interests over ours"; "This is an attempt to eliminate the leading American drone company and deepen the world's dependence on Chinese drone suppliers."

2. South China Morning Post. "The Story of Drone Pioneer DJI." September 20, 2019. DJI formally established in Shenzhen in 2006; early prototypes developed by Frank Wang while a student at HKUST. Also: Fortune (August 2017); Frank Wang biographical sources confirming Shenzhen registration.

3. Loyalty Drone Research. U.S. Commercial Drone Market Share Analysis. 2025. DJI holds approximately 80 percent of the U.S. commercial drone market.

4. Dedrone. State of Drone Security Report. 2025. DJI platforms accounted for 83.48 percent of all drone detections globally in 2025, across conflict zones and operational theaters.

5. DJI-commissioned economic analysis. The Economic Impact of DJI Products and Services in the United States. Published 2023–2024. DJI products and services estimated to contribute $116 billion in economic value and support nearly 450,000

U.S. jobs. Note: figure derived from DJI-commissioned analysis; methodology not independently verified.

6. Foundation for American Innovation. The Drone Dependency: Chinese UAS Procurement by U.S. State Agencies. 2023. From 2010 to 2022, on average 85 percent of drones purchased by U.S. state agencies were Chinese-made.

7. United States Congress. James M. Inhofe National Defense Authorization Act for Fiscal Year 2024. Public Law 118-31. Signed December 22, 2023. Includes the American Security Drone Act, prohibiting federal funds for covered foreign-entity UAS procurement.

8. Federal Acquisition Regulation interim rule. Published in the Federal Register. Effective November 12, 2024. DoD, GSA, and NASA prohibited from procuring covered UAS in contracts awarded on or after that date.

9. Office of Management and Budget. Government-Wide Implementation Memorandum: Secure UAS Procurement Requirements. November 21, 2025. Extends secure UAS procurement requirements to all federal agencies and all recipients of federal funds.

10. Federal Communications Commission. National Security Covered List. Updated December 2025. DJI added following FY2025 NDAA review requirement; automatic listing triggered by failure to complete review before statutory deadline. DJI filed suit against FCC, February 2026; 25 products and approximately $1.5 billion in planned U.S. revenue frozen. Also: Defense Innovation Unit, Blue UAS Cleared List, updated August 2025; 23 drone platforms and 14 UAS framework components; 369 proposals from 19 nations in 2025 Refresh Challenge.

11. Skydio, Inc. SkyForge Initiative and Series F Funding Announcement. April 2026. $3.5 billion five-year investment; manufacturing facility five times current size; 2,000+ direct jobs; 3,000+ supply chain jobs; $1 billion+ to domestic suppliers; Series F at $4.4 billion valuation; 60,000+ drones delivered to 3,800+ customers including 1,200+ public safety agencies, all U.S. military branches, and 29 allied nations.

12. The Next Web. "U.S. Banned DJI but China Controls 90% of Rare Earth Magnets and 99% of Drone Batteries Needed to Replace It." May 2026. China controls approximately 90 percent of global rare earth mineral processing, approximately 99 percent of drone battery cell production, and approximately 90 percent of permanent magnets that power drone motors. Corroborated by: The Diplomat (April 2026); International Energy Agency data (91 percent rare earth separation and refining, 94 percent sintered permanent magnet production, 2024).

13. White House. Executive Order 14307: "Unleashing American Drone Domination." Signed June 6, 2025. Also: U.S. Department of Defense. Secretary of Defense Pete Hegseth, memorandum "Unleashing U.S. Military Drone Dominance." July 10, 2025. Rescinds 2021 and 2022 Biden-era drone procurement

restrictions; delegates procurement authority to combat units; directs DIU to scale the Blue UAS Cleared List.

14. OnDefend. DJI Technology Inc. Security Assessment Executive Report. Published May 2026. Five-month independent cybersecurity assessment of DJI Air 3S and DJI Matrice 4E platforms. Methodology included software and firmware analysis, hardware teardown, RF spectrum analysis (1 MHz–6 GHz), network traffic monitoring, supply chain verification, adversarial attack simulations, replay and jamming testing, and jailbreak attempts. Findings: no evidence of hidden backdoors; no unauthorized data transmission outside the United States; no viable pathways for hijacking or weaponizing the tested platforms; DJI Local Data Mode confirmed to prevent transmission of flight data to internet-based destinations while enabled; no evidence of deferred data upload upon return to normal operating mode. Ten low-risk findings and thirteen observations identified, none representing realistic threats to flight safety or widespread data exposure. Published by Drone Service Providers Alliance, May 28, 2026.

15. Association for Uncrewed Vehicle Systems International (AUVSI). Green UAS Program Overview. 2025. Green UAS provides cybersecurity and supply-chain vetting for non-DoD operators including state, local, commercial, and critical infrastructure users. Designed to bridge the gap between NDAA compliance as a legal threshold and Blue UAS as a DoD-specific certification pathway.

The Talent Deficit

Why the Industry Can't Grow Without People

"Uncrewed aviation will transform the world but only if we prepare the people who will transform the industry."

— Sharon Rossmark

The Contracts Nobody Could Fill

On a Tuesday morning in the spring of 2025, a workforce development director at a regional drone services firm sat at her desk looking at a spreadsheet she had been staring at, with variations, for the better part of two years. One column listed active client inquiries: utilities wanting infrastructure inspection, county agencies requesting emergency management surveys, agricultural cooperatives asking about crop health monitoring, a logistics firm exploring last-mile delivery pilots. The adjacent column listed available, qualified personnel capable of executing those contracts. The gap between the two columns was not closing. If anything, it was widening.

The phones were ringing. The interest was real. The companies requesting bids had budget, they had urgency, and they had executive mandates to integrate uncrewed systems into their operations. The regulatory environment, while still evolving, had opened enough doors to make commercial drone operations a practical reality across dozens of use cases. And the equipment, the aircraft themselves, charged and ready, was sitting in storage cases on metal shelves in a climate-controlled back room, waiting.

What was missing were the people. Not enough qualified remote pilots to staff concurrent missions. Not enough trained technicians to maintain a growing fleet with the rigor and reliability clients expected. Not enough operations managers who understood both the regulatory landscape and the logistics of coordinating multi-aircraft missions across wide geographic areas. The firm had turned away contracts worth hundreds of thousands of dollars. Not once, but repeatedly. It had done so quietly, the way companies absorb operational failures they cannot fully explain to clients, with regret and without public announcement.

Her experience was not an anomaly. Talk to enough workforce development professionals, UAS operations directors, and regional aviation educators, and a pattern emerges that is consistent enough to be called a structural condition. The drone industry in the United States has arrived, with striking speed, at a moment of genuine market opportunity. The technology works. The applications are proliferating. The regulatory path, however slow and contested, is clearing. Investment is flowing. And at the very intersection of readiness and growth, the industry has encountered the one constraint it did not fully anticipate and has not yet solved: the absence of a trained, credentialed, and sufficiently diverse workforce in the numbers the market requires.

In an industry defined by technological innovation, by the relentless pace at which the hardware improves, the software matures, and the operational envelopes expand, how did human capital become the single greatest constraint on growth?

More Than a Pilot Shortage

The instinct, when people discuss the UAS workforce shortage, is to reach for the pilot numbers. And in isolation, those numbers are remarkable. As of December 31, 2025, the Federal Aviation Administration reported 492,311 active Part 107 Remote Pilot certificates across the United States, nearly half a million certificated operators trained in the time it might take a traditional aviation program to graduate a few thousand.[1] Viewed through a certain lens, this sounds like a workforce success story. It is not the whole story.

Those 492,311 certificates were earned by people who successfully passed the FAA Aeronautical Knowledge Test, which qualified them to legally pilot a commercial drone within a specific operational envelope: visual line of sight, one operator managing one aircraft at a time, missions of bounded scope and duration. What the Federal Aviation Administration's long-anticipated Part 108 rules for Beyond Visual Line of Sight (BVLOS) operations will demand is a fundamentally different kind of workforce doing a fundamentally different kind of work.

"Instead of one aircraft to one pilot, you have one pilot to tens, if not hundreds, of aircraft at once, acting more like an air traffic controller rather than a pilot."

That characterization comes from James McDanolds, Program Chair of the School of Uncrewed Technology at Sonoran Desert Institute, speaking to Aviation International News in April 2026.[3] It is a precise summary of an operational transformation that changes not just the job description of a drone pilot, but the entire organizational structure required to support drone operations. McDanolds went further, estimating that within two to three years of the FAA finalizing Part 108, the U.S. aviation industry could need hundreds of thousands of additional workers, not pilots alone, but the full occupational ecosystem required to sustain large-scale, long-range, and heavily automated drone operations.

That ecosystem is both broader and more technically demanding than the public conversation about drone jobs typically acknowledges. Remote pilots are only the most visible layer. Behind every operational flight, someone performs the work that keeps the mission running: maintaining the aircraft, processing the sensor data, coordinating the airspace authorization, managing the logistics. In small operations, which make up the majority of the commercial drone sector, those functions often fall to the same one or two people. In larger enterprise deployments, they require dedicated professionals: technicians, data analysts, operations managers, fleet managers, regulatory affairs specialists, software engineers, and logistics coordinators. Whether those functions are performed by a single versatile operator or by a team of specialists, they all require trained people to execute them. The drone industry does not have a pilot shortage. It has a talent bottleneck across an entire occupational ecosystem.

Among all of these gaps, the technician shortage is simultaneously the most acute and the least publicly discussed. This is partly because the technician problem is invisible in a way the pilot problem is not. When there are not enough pilots, contracts do not get signed. When there are not enough technicians, aircraft ground unexpectedly, missions abort mid-sequence, and the reliability numbers that enterprise clients track begin to erode. But there is a second, deeper reason the technician problem has escaped adequate attention: there is, quite simply, no regulatory baseline for drone technician qualifications in the United States.[11]

Nothing analogous to the FAA's Airframe and Powerplant (A&P) certificate, the hard-won, federally standardized credential that has defined what it means to be a qualified aircraft mechanic for more than half a century, has been established for the UAS maintenance workforce. The A&P framework took decades to build, reflects thousands of hours of required instruction, and carries the weight of federal law. Its equivalent for drones does not yet exist. What drone technician roles actually demand is technically sophisticated: electrical hardware troubleshooting at the component level; systems-level knowledge of flight controllers, electronic speed controllers, and the motors that translate digital commands into physical motion; the ability to interpret autopilot flight data logs to diagnose faults before they become failures; and familiarity with the increasingly varied materials from which modern drone airframes are built, including carbon-fiber composites and components produced through additive manufacturing. Companies are hiring for this work daily. They are doing so against no common standard, with no established credential, and with a pool of candidates who largely learned drone maintenance not through any formal program, but by acquiring skills in adjacent fields and applying them, through practice and necessity, to uncrewed systems.

145 Schools and Counting: The Education System Plays Catch-Up

The clearest way to understand how hard the American education system has worked to build a UAS workforce pipeline, and how far it still has to go, is to trace the arc of the FAA's UAS Collegiate Training Initiative (UAS-CTI). The UAS-CTI program launched on April 30, 2020, with a founding cohort of 26 schools. Five years later, by January 2025, the program had grown to more than 145 institutions, according to the FAA's own five-year program review: technical colleges, community colleges, and four-year universities spanning every region of the country.[4]

The growth curve is genuinely impressive. What it cannot disguise, however, is how low the starting point was, and how the expansion of the program still lags the expansion of the industry it is meant to serve. More than 145 schools is a meaningful number. It is not, by any reasonable analysis, enough. The commercial drone market is projected to grow more than 50 percent in the next five years, and the operational demands of Part 108 BVLOS rules will multiply the per-flight workforce requirements at precisely the moment the market accelerates. The pipeline is growing. The gap it is meant to close is growing faster.

The experience of West Georgia Technical College offers a close-range view of how this dynamic plays out in practice. In September 2025, West Georgia Tech launched what was described as the first drone training program of its kind in Georgia, a state that was, at that same moment, emerging as a national leader in agricultural drone adoption.[6] Georgia farmers were already using drones to monitor crop health, apply targeted inputs, and survey acreage at a speed and resolution that traditional methods could not match. The technology was in the field. The business case had been made. And yet, until the fall of 2025, no institution in the state offered a formal technician training program aligned to the needs of that industry. These two facts, Georgia as a drone agriculture leader, Georgia with no formal drone technician training until 2025, were simultaneously true, and their coexistence is precisely the pattern that characterizes the national landscape. Industries adopt; education follows. The gap between adoption and training is where talent shortages live, and they tend to persist until someone deliberately builds the bridge.

Industries adopt; education follows. The gap between adoption and training is where talent shortages live, and they tend to persist until someone deliberately builds the bridge.

The Unmanned Safety Institute (USI) moved to build that bridge in February 2026, launching UAS Maintenance Technician programs at three community colleges simultaneously: Central Arizona College, Horry-Georgetown Technical College in South Carolina, and West Georgia Technical College.[5] The design of these programs reflects a deliberately accelerated approach: twelve weeks, aligned to ASTM International F3600 standards, the technical benchmark increasingly recognized by employers and procurement officers as the reference point for drone safety and maintenance competency. Graduates earn the USI UAS Maintenance Technician Designation, a credential that employers can evaluate against a defined standard rather than the informal, inconsistent experiential claims that currently dominate the hiring process for drone maintenance roles. These are not traditional four-year degrees, and they are not intended to be. They are workforce pipelines designed to move a motivated student from enrollment to employability in a single semester, reaching populations that four-year aviation programs do not.

The broader aerospace and defense workforce picture provides context that qualifies even the most encouraging program announcements. The Aerospace Industries Association (AIA), in partnership with McKinsey, released its 2025 Aerospace and Defense (A&D) Workforce Study in June of that year.[7] The findings were stark. The industry was growing at 4.8 percent per year while simultaneously suffering an attrition rate of nearly 15 percent. Seventy-six percent of AIA member organizations reported sustained challenges hiring engineering talent. Fifty-six percent reported difficulties sourcing skilled trades workers.

The Aviation Technician Education Council (ATEC), working with the Oliver Wyman aviation analytics team, found in its September 2025 pipeline report that certificated mechanics were already in shortage by roughly 10 percent in 2025, a gap projected to narrow only partially, to 7 percent, by 2035.[8] Even at that narrowed gap, the shortfall would represent approximately 10,000 missing certificated mechanics, and that projection was made before a single UAS-specific role was added to the demand calculation. The FAA issued slightly more than 9,000 new mechanic certificates in all of 2024. Aviation maintenance technician school enrollment increased 9 percent, encouraging, but nowhere near sufficient to meet the compounding demand that BVLOS operations will generate.

Half the Talent Pool, Mostly Untapped

There is a dimension of the UAS talent deficit that the workforce pipeline data alone does not fully capture, and it is perhaps the most consequential of all. It has to do not with how many people the industry is training, but with who those people are, and, more precisely, who they are not.

Of the 492,311 active Part 107 Remote Pilot certificates recorded by the FAA as of December 31, 2025, women hold 44,274, roughly 9 percent of the total certificated remote pilot workforce. That 9 percent grew by 19.3 percent year-over-year in 2025, a meaningful acceleration driven in part by the sustained visibility and community-building work of organizations focused on expanding access to UAS careers for workers not yet reached by the industry's traditional pipelines.[2] The geographic distribution of that growth is concentrated in agricultural heartland states rather than the coastal technology hubs most commonly associated with drone innovation—a pattern that Chapter 6 examines in full. The average woman entering the commercial drone pilot workforce in 2025 was approximately 39 years old: a mid-career professional making a deliberate and considered pivot, not a recent graduate following a well-lit path.

The 19.3 percent growth rate is genuinely encouraging. The 9 percent total share is genuinely alarming. An industry that is simultaneously facing a critical workforce shortage while drawing from only 9 percent of half the available working-age population is not making a social calculation. It is making a strategic error. The failure to recruit, train, and retain women in the UAS workforce is not primarily a participation gap in the narrow sense of that phrase as it is commonly used in corporate settings. It is a capacity problem. An industry that cannot fill its talent gap and simultaneously refuses, by design or by default, to recruit from 50 percent of the available workforce has constructed its own crisis.

The gap reflects structural challenges that go beyond the entry credential—encompassing cultural barriers, mentorship deficits, and retention failures that Chapter 6 examines in full, with the research and policy implications that a landscape chapter can only introduce.

The talent gap extends well beyond gender. Rural communities, post-industrial towns, and lower-income urban neighborhoods are largely disconnected from the UAS career pathway for reasons that are structural and solvable—and that Chapter 7 addresses in full.

The Price of the Gap

Workforce shortages have a way of remaining abstract until they are translated into the currency most familiar to business: lost revenue, foregone opportunity, and competitive disadvantage that compounds over time. The UAS talent deficit is no longer abstract. It is costing the industry money, mission capability, and strategic position in a global competition the United States cannot afford to lose.

Every unfilled UAS position represents a contract that was not signed, a mission that was not flown, a capability that went undeployed. The infrastructure inspection company that cannot hire enough qualified operators turns away utility clients who have capital budgets, regulatory mandates, and operational urgency. The precision agriculture firm that cannot staff its Midwest operations leaves farmers without the crop health data that would allow them to apply inputs more precisely, reduce waste, and improve yields in a season where commodity margins are thin. The public safety agency that cannot staff its drone program relies instead on methods that are slower, more expensive, and less safe. These are not hypothetical costs. They are the documented experience of operators across the industry, repeated in conversation after conversation with workforce development professionals and company executives who describe turning away business as routine operational reality rather than exceptional circumstance.

The cost compounds at the national level in ways that are increasingly difficult to separate from the larger geopolitical context of drone policy. The United States has made an explicit, policy-driven commitment to rebuilding domestic drone manufacturing and reducing federal procurement dependence on Chinese-made systems. The National Defense Authorization Act (NDAA) restrictions have created the demand signal. The Blue UAS framework has identified and approved domestically manufactured alternatives. The manufacturing ambitions described in the previous chapter, Skydio's SkyForge commitment among them, represent the supply-side response to that policy signal. But a factory that is five times larger than its predecessor requires five times as many workers: engineers, precision manufacturing technicians, quality control specialists, supply chain managers, and assembly line workers trained to the tolerances and documentation standards that defense and enterprise customers require. The manufacturing imperative cannot be executed without the workforce this chapter describes. These are not parallel problems that can be addressed in sequence. They are the same problem, viewed from two different vantage points on the same factory floor.

The global competition for UAS talent adds a further dimension to the urgency. China, which has invested aggressively and systematically in drone manufacturing capacity and international market penetration, has invested with equal deliberateness in the technical and engineering workforce that sustains it. DJI alone employs thousands of engineers at its Shenzhen headquarters, supported by a national education and industrial policy that has prioritized uncrewed systems as a strategic sector for decades. The global drone industry now employs more than 2.1 million people worldwide.[10] The United States is building its UAS workforce pipeline at the precise moment the industry needs it, rather than in anticipation of that need. That sequence has costs that show up not in this quarter's earnings reports, but in the decade-long competitive positioning of American drone manufacturing and services relative to competitors who started earlier and invested more deliberately.

The Runway Is Short

The talent deficit in uncrewed aviation is serious, urgent, and consequential. It is also solvable. The evidence for that claim is not wishful thinking. It is observable in the programs that are already working, in the enrollment numbers that are already rising, and in the workforce dynamics that have already begun to shift. The question is not whether the gap can be closed. It is whether the closing will happen fast enough to match the pace of the industry's expansion.

The UAS-CTI program's growth from 26 founding schools in April 2020 to more than 145 institutions by January 2025 proves that the American education system can move with meaningful speed when it is given the right incentives, the right frameworks, and the right partners in industry and government.[4] Five years and 119 additional institutions is not a slow bureaucratic response. It is a genuine mobilization, accomplished during a period that included a global pandemic, significant community college enrollment disruptions, and the ongoing regulatory uncertainty that makes it difficult for institutions to build programs around a regulatory target that is still moving. The USI's 12-week maintenance technician programs prove that high-quality, workforce-ready credentials do not require four-year degree timelines. The 19.3 percent year-over-year growth in women earning Part 107 certificates in 2025 proves that the untapped talent pool will respond when the opportunity is made visible, accessible, and explicitly welcoming.

The Part 108 NPRM, published on August 7, 2025, drew more than 3,000 public comments, a level of engagement that reflects the depth of economic interest in what the BVLOS framework will enable.[9] With implementation expected six to twelve months after the final rule is published, real-world operational impact is likely to begin arriving in late 2026 or early 2027. That is not a distant horizon. It is a runway, and it is short.

What is missing is not the will of individual institutions, individual instructors, or individual companies. What is missing is coordination: a national workforce strategy for uncrewed aviation that connects the FAA's CTI schools with the employers who need their graduates; that aligns Department of Labor workforce development funding with the credential pathways the industry actually values and can use; that gives community colleges the resources to build programs before demand becomes a crisis rather than in response to one; and that actively and intentionally recruits from the communities that represent the largest available expansion of the UAS talent base. Women, whose participation remains far below what the industry's long-term capacity requires. Communities in rural and post-industrial regions that the industry has not yet meaningfully engaged, despite their proximity to emerging aviation pathways. Rural youth, for whom drone agriculture is not an abstraction but a daily reality and a natural point of entry into UAS careers. And veterans, a workforce whose technical training, operational discipline, and mission orientation make them among the most readily translatable talent assets the industry has access to.

Technology is not the bottleneck. The market is not the bottleneck. The regulatory path is clearing, contract by contract and rulemaking by rulemaking. The workforce is the bottleneck. And it is, of all the constraints currently shaping the trajectory of American uncrewed aviation, the one most directly within the country's own control. Capital flows where it will. Adversary nations manage their own supply chains. But the United States can choose, as a matter of deliberate national policy and coordinated industry investment, who it trains, how it trains them, and how fast. That choice, and the structures, incentives, and institutions required to make it well, is the subject of Part II of this book.

Looking Ahead — Chapter Four

Chapter Four takes the argument to the factory floor. The workforce crisis and the manufacturing imperative are not parallel problems. They are one problem. Building

American drones requires American workers, from the engineers who design the systems to the technicians who assemble the airframes to the quality control specialists who certify the components. That is where the story goes next.

Notes and Sources

1. Federal Aviation Administration. U.S. Civil Airmen Statistics, December 31, 2025. FAA Aeronautical Center, Oklahoma City, OK. Active Part 107 Remote Pilot Certificates: 492,311.

2. Federal Aviation Administration. Civil Airmen Statistics, December 31, 2025. FAA Aeronautical Center, Oklahoma City, OK. Women holding active Part 107 Remote Pilot Certificates as of December 31, 2025: 44,274 (approximately 9% of total certificated remote pilot workforce). Year-over-year growth: 19.3%. States with highest growth rates: South Dakota (27.3%), Indiana (26.6%), Wisconsin (25.1%). Average age of women entering the field: approximately 39.

3. McDanolds, James (Program Chair, School of Uncrewed Technology, Sonoran Desert Institute). Aviation International News, April 21, 2026. On the operational shift under Part 108 BVLOS rules and projected workforce demand within two to three years of rule finalization.

4. Federal Aviation Administration. UAS Collegiate Training Initiative Five-Year Review, 2020–2025. April 2025. Program launched April 30, 2020, with 26 founding schools. Grown to more than 145 institutions by January 2025.

5. Unmanned Safety Institute (USI), press release, February 2026. Launch of UAS Maintenance Technician programs at Central Arizona College, Horry-Georgetown Technical College (SC), and West Georgia Technical College. Twelve-week accelerated curriculum aligned to ASTM International F3600 standards. Graduates earn the USI UAS Maintenance Technician Designation.

6. West Georgia Technical College / Unmanned Safety Institute, announcement, September 2025. Launch of WGTC's inaugural drone training program in partnership with USI; described as the first of its kind in Georgia.

7. Aerospace Industries Association (AIA) and McKinsey & Company. 2025 Aerospace & Defense Workforce Study. June 2025. Industry annual growth rate: 4.8%; attrition rate: approximately 15%; 76% of AIA members report sustained engineering hiring challenges; 56% report skilled trades sourcing challenges.

8. Aviation Technician Education Council (ATEC) and Oliver Wyman. Vector: 2025 Aviation Technician Pipeline Report. September 2025. Certificated mechanics in shortage by 10% in 2025; projected 7% shortfall (approximately 10,000 mechanics) by 2035. FAA issued slightly more than 9,000 new mechanic certificates

in 2024. Aviation maintenance technician school enrollment increased 9%. U.S. civil commercial aviation maintenance workforce: more than 431,000 personnel.

9. Federal Aviation Administration. Notice of Proposed Rulemaking: Operation of Unmanned Aircraft Systems Beyond Visual Line of Sight (Part 108). Published August 7, 2025. Comment period closed October 6, 2025. Approximately 3,100 public comments received (source: FAA Federal Register, January 28, 2026; DLA Piper, October 2025). Final rule implementation expected 6–12 months after publication.

10. StartUs Insights. Global Drone Industry Report. 2025. Global drone industry estimated employment: more than 2.1 million people worldwide.

11. Aviation International News (AIN), April 2026. Analysis of UAS technician qualification gap; description of drone maintenance competency requirements including electrical hardware troubleshooting, flight controller systems knowledge, autopilot data interpretation, and composite/additive airframe material familiarity. Absence of regulatory baseline analogous to the FAA Airframe and Powerplant (A&P) certificate for UAS maintenance roles.

12. Multiple industry sources, 2025–2026. Commercial drone market projected growth of more than 50% over the next five years. Sources include Drone Industry Insights, PwC Clarity from Above (updated analysis), and Goldman Sachs Research.

The Manufacturing Imperative: American Drones, American Jobs

"The future of uncrewed aviation will belong to the nations that choose to build, not just buy."

— Sharon Rossmark

The Factory That Hasn't Been Built Yet

The building sits on the edge of a mid-sized city in Ohio, not far from the interstate that once carried American-made automobiles to every corner of the country. It is a hundred thousand square feet of precisely climate-controlled floor space, the kind of facility that manufacturers spent decades perfecting for tolerances measured in thousandths of an inch. Computer Numerical Control (CNC) machining centers line one wall. Precision assembly stations occupy the center. A quality assurance lab anchors the far end. The workers who report here every morning are experienced in aerospace-grade fabrication, people who understand the difference between a part that is good and a part that is certified. The facility has capacity. It has capability. It has a workforce. What it has not had, until very recently, is an unambiguous signal from Washington that the product its management team has been eyeing for three years is worth the risk of retooling.

That product is drones. And for the better part of a decade, the economic and regulatory logic simply did not close. Chinese manufacturers, led by DJI, had built a global supply chain of extraordinary efficiency, producing commercial uncrewed aerial systems at price points that no American factory could match while also absorbing the regulatory uncertainty that had kept U.S. commercial drone markets from scaling. The facility in Ohio watched, as did dozens like it across the industrial Midwest. Management attended drone industry conferences. Engineers flew DJI platforms to understand the technology. The finance team ran pro formas on drone production economics and put them back in the drawer. The policy signal was not clear enough. The demand was not guaranteed enough. The capital risk was too great.

Then came the summer of 2025. An executive order. A landmark piece of legislation. A Pentagon procurement directive. A regulatory agency expanding its list of prohibited foreign technology. In a span of roughly eight weeks, the calculus that had kept American manufacturers on the sidelines of drone production was rewritten. The phone at the Ohio facility began to ring. Conversations that had been theoretical for years became operational almost overnight. Procurement officers. Defense contractors. Systems integrators. Each one asking a version of the same question: can you make drones, and how fast can you start?

The question that had long seemed rhetorical, can American manufacturers build drones at scale, suddenly had a deadline attached to it. And the answer, it turns out, is complicated in exactly the ways that matter most.

American drone manufacturing is not a distant aspiration. It is an urgent national project, backed by billions in federal investment and driven by genuine geopolitical necessity. But building a domestic drone manufacturing base requires confronting a set of interdependent challenges, supply chain sovereignty, workforce capacity, cost competitiveness, and policy consistency, that no single executive order or legislative act can resolve on its own. The story of how America is attempting to meet that challenge is a story about industrial ambition, strategic vulnerability, and the indispensable human capital without which every other element of the manufacturing imperative is simply machinery waiting for someone qualified to run it.

The Summer of 2025

Historians of industrial policy tend to look for moments when the government's intent becomes not merely stated but financially legible, when the words in a document are matched by appropriated dollars and enforceable directives. The summer of 2025 was such a moment for American drone manufacturing. What unfolded between June and August of that year was not a single policy decision but a coordinated sequence of actions that, taken together, represented the most serious federal commitment to domestic uncrewed systems production in the industry's history.

It began on June 6, 2025, when President Trump signed the executive order titled "Unleashing American Drone Dominance." The title alone signaled something new. Previous federal actions on drone policy had been couched in the careful language of safety, integration, and regulatory clarity. This order used a different vocabulary. "Dominance" is a word from geopolitics and industrial strategy, a declaration of ambition, not merely a statement of preference. The order directed the Federal Aviation Administration (FAA) to issue a proposed rule on Beyond Visual Line of Sight (BVLOS) operations within thirty days, a timeline that, by FAA standards, was not a target so much as a mandate.[1] It called for streamlining certification and approval processes. It directed federal agencies to prioritize procurement of American-made uncrewed aerial systems. And it called explicitly for expanding the export of trusted, American-manufactured UAS to global markets, signaling not just domestic substitution but international commercial competition.

The order used a different vocabulary. "Dominance" is a word from geopolitics and industrial strategy — a declaration of ambition, not merely a statement of preference.

Then came Independence Day. The choice to sign the One Big Beautiful Bill Act (OBBBA) on July 4, 2025, was not a coincidence of legislative timing.[2] The symbolism was deliberate, and the substance was extraordinary. The OBBBA's drone and robotics provisions went beyond anything the industry had previously seen from a single piece of legislation: \$1.4 billion to scale the small UAS industrial base; \$2.1 billion for medium unmanned surface vessel development; \$188 million for maritime robotic autonomous systems research and development; and \$250 million for a digital test environment designed specifically for autonomy validation. These were not research grants or pilot programs. They were industrial investment commitments at a scale that told drone manufacturers, component suppliers, and venture investors something they had been waiting years to hear: the federal government is prepared to be the anchor customer for a domestic drone manufacturing renaissance, and it has put real money behind that commitment.

The Department of Defense (DoD) procurement directive of July 2025 completed the trifecta. Secretary Pete Hegseth signed an order accelerating purchase timelines and expanding budgets specifically to enable the military branches to rapidly deploy domestically produced drone platforms. Combined with the Department of Government Efficiency's (DOGE) initiative at the Pentagon, consolidating drone procurement orders from multiple service branches and targeting the purchase of at least 30,000 small, low-cost drones under a single coordinating framework, the message to the private sector was unmistakable.[3] The U.S. government is buying American drones. In large quantities. With real money. On an accelerated timeline.

Private capital did not wait for further clarification. From May through July 2025, U.S. drone and robotics companies raised hundreds of millions of dollars in fresh capital, with investment announcements consistently citing the legislative and executive signal as the driving catalyst.[4] Firestorm Labs closed a $47 million Series A to accelerate modular domestic drone production for federal contracts. Quantum Systems raised €160 million in a Series C to scale dual-use autonomous aircraft technologies. Unusual Machines saw its stock surge 40 percent on the strength of the Pentagon procurement announcements. The AIRO Group completed a public offering in June, bringing capital-markets access to a sector that had long relied on private funding. The investment wave of that summer was not coincidental enthusiasm. It was the rational response of a capital market reading a credible demand signal from the most consequential buyer in the economy.

The sequence, executive order, OBBBA, DoD directive, DOGE consolidation, represents the most coherent and financially serious federal commitment to domestic drone manufacturing in the industry's history. Whether it will be sufficient depends entirely on what happens next, on the factory floor and in the supply chain that runs beneath it.

Sixty Companies and Counting

The American drone manufacturing ecosystem is larger, more technically sophisticated, and more commercially serious than most observers outside the industry recognize. According to Bolt Flight's comprehensive industry survey published in May 2025, more than sixty U.S. drone manufacturers are now actively producing uncrewed aerial systems, spanning compact inspection UAVs to heavy-lift autonomous platforms, tactical intelligence and surveillance systems to agricultural sprayers, infrastructure inspection tools to public safety platforms.[5] The picture that emerges from that survey is of an industry that has been quietly building capability for years while the policy environment slowly caught up to its ambitions.

At the front of the formation is Skydio. Based in San Mateo, California, Skydio made a consequential strategic decision in 2023 when it exited the consumer drone market to concentrate entirely on enterprise and defense. At the time, that choice looked like a retreat. In retrospect, it looks like one of the most prescient industrial pivots of the decade. With more than 60,000 drones delivered to 3,800-plus customers, including every branch of the U.S. military and twenty-nine allied nations, Skydio is not a startup with a manufacturing plan. It is an established defense and enterprise supplier with a scale-up strategy of breathtaking ambition.[6]

That ambition was made explicit in April 2026 with the announcement of the SkyForge program, a $3.5 billion manufacturing expansion initiative that, if fully executed, would represent the largest single manufacturing commitment in U.S. drone industry history. The SkyForge plan calls for a new production facility five times larger than Skydio's current operations, more than 2,000 direct manufacturing jobs, more than 3,000 additional positions in the supply chain, and more than $1 billion directed specifically to domestic component suppliers. The accompanying Series F funding round, $110 million at a $4.4 billion valuation, signals that institutional investors regard this ambition as financially viable, not merely rhetorically bold. SkyForge is what it looks like when a drone company decides to build not just a factory, but an industrial ecosystem.

Behind Skydio, a diverse cohort of manufacturers has been building for specific niches within the defense, public safety, and critical infrastructure markets. AeroVironment, one of the most established names in U.S. military uncrewed systems, produces the Raven and Puma platforms that have defined small intelligence, surveillance, and reconnaissance (ISR) drone operations for the American military for more than a decade. Teal Drones, acquired by Red Cat Holdings, has built a reputation for security-hardened platforms optimized for contested environments. Inspired Flight, Freefly Systems, ModalAI, Teledyne FLIR, Shield AI, Wingtra, and Parrot round out the Defense Innovation Unit's (DIU) Blue UAS Cleared List, a catalogue of twenty-three drone platforms and fourteen UAS framework components that have been vetted for use in sensitive government applications.[7] The cleared list is, in effect, the federal government's endorsement of a domestic drone supply base: these companies, their products, and their supply chains meet the security bar that the national security enterprise requires.

But intellectual honesty demands a candid accounting of what this ecosystem is, and what it is not yet. It is a collection of genuine companies building real products for real customers in markets where security requirements, not price, drive procurement decisions. It is not yet a manufacturing base of the scale, depth, and cost competitiveness that can replace DJI across the full commercial marketplace. In the broader commercial economy, agricultural spray drones, entry-level survey systems, real estate photography platforms, consumer inspection tools, there is still no domestically manufactured alternative that competes on price with Chinese-made platforms. The Blue UAS cleared companies are not, for the most part, trying to win the mass-market segment. They are building for government, public safety, and critical infrastructure, markets defined by the willingness to pay a security premium. Transforming from a cleared defense supplier into a full-spectrum commercial incumbent is a journey measured in years and billions of dollars of sustained investment. The cleared ecosystem is the foundation. It is not yet the edifice.

What "Made in America" Actually Means

The Chinese government's October 2024 decision to shut down Skydio's battery supply chain, documented in Chapter 2, demonstrated something that procurement policy as currently constructed has not yet fully absorbed: the location of the final assembly line is not the same thing as manufacturing sovereignty. CEO Adam Bry's response captured the principle precisely: the company rebuilt what he described as "by far the most secure drone supply chain in the world, independent from China." But Skydio's survival required extraordinary effort to route around a single component dependency. Most American drone manufacturers have not yet made that effort, and the legislative framework does not yet require them to.

A drone assembled in an American factory from a Chinese battery, Chinese sensors, Chinese motors, and a Chinese-origin software architecture is not a domestically sovereign drone. It is a Chinese drone with an American street address.

The underlying data makes this vulnerability concrete and unmistakable. China controls approximately 90 percent of global rare earth mineral processing, the materials essential to the sensors, motors, and electronic components that make a drone functional, regardless of where it is finally assembled.[8] As documented in Chapter 2, China produces an estimated 99 percent of drone battery cells globally and approximately 90 percent of the permanent magnets that power drone motors. These are not incidental supply chain inconveniences that can be resolved with a procurement policy revision. They are structural dependencies baked into the global industrial architecture over decades of investment, scale, and deliberate strategic cultivation.

This is the "assembled in America" problem, and it explains why the American Security Drone Act (ASDA), enacted on December 22, 2023, as part of the National Defense Authorization Act (NDAA) for fiscal year 2024, and reinforced by the Federal Acquisition Regulation (FAR) interim rule that took effect on November 12, 2024, is a necessary but insufficient solution.[9][10] The ASDA prohibits the use of federal funds to procure or operate drones from covered foreign entities. The FAR rule closes the contracting loophole by prohibiting DoD, the General Services Administration (GSA), and the National Aeronautics and Space Administration (NASA) from procuring UAS from ASDA-covered foreign manufacturers in any contract awarded on or after that date. These are meaningful restrictions. They change the nameplate on the factory. They do not yet change the geography of the supply chain that feeds it. And the Federal Communications Commission's (FCC) December 2025 expansion of its Covered List, explicitly including DJI and signaling a broader direction toward all foreign-manufactured UAS and critical UAS components, suggests that regulators are beginning to grapple with the component dimension of the problem as well.[11]

Full manufacturing sovereignty requires something the current legislative framework has only begun to address: component sovereignty, meaning domestic or trusted-ally sourcing at every meaningful level of the bill of materials. The Skydio SkyForge initiative represents the most serious attempt yet to confront this reality directly. By directing more than $1 billion specifically to domestic component suppliers, and by inviting those suppliers to co-locate with Skydio's manufacturing operations, Skydio is attempting to rebuild the supply chain infrastructure itself, not merely optimize the final assembly line. This is the architectural distinction that separates a genuine domestic drone industry from a sophisticated repackaging operation. Manufacturing sovereignty is not a factory. It is an ecosystem, from raw materials through finished product, with domestic or trusted-ally sourcing at every critical link. The broader U.S. drone sector is at the very beginning of that journey.

The Demand Signal Is Not Enough

The Replicator program was, in many ways, the proof of concept that proved the problem. Launched in 2023 with genuine urgency and serious high-level backing, Replicator set out to demonstrate that the United States military could field thousands of attritable autonomous systems at speed and scale, matching, in concept if not in quantity, the drone production rates that America's adversaries had demonstrated were achievable. The program succeeded in surfacing and articulating the operational requirements. It succeeded in generating political momentum and directing attention to the scale problem. What it struggled to do was translate that momentum into scaled domestic production.[3]

The structural reason was not a failure of intent. It was a classic industrial policy dilemma, the same dynamic that plays out whenever a government attempts to stimulate private manufacturing investment through signals rather than commitments. Domestic drone manufacturers couldn't justify the capital expenditure required to scale production capacity without firm, large-volume purchase contracts. The government couldn't commit to firm, large-volume contracts without confidence that domestic manufacturers could actually deliver at scale. Each party was waiting for the other to move first. Replicator identified the operational imperative with clarity. It could not resolve the financing logic that kept manufacturers on the sideline.

This is the structural problem that the semiconductor industry confronted a generation earlier, and that the Creating Helpful Incentives to Produce Semiconductors (CHIPS) and Science Act of 2022 was specifically designed to break. The solution in semiconductors was not to ask the private sector to take the full risk of re-establishing domestic manufacturing in a market dominated by far-more-efficient foreign competitors. It was to use federal purchasing commitments, subsidies, and tax incentives to reduce the cost and risk of domestic scale-up to the point where private investment became viable. The drone manufacturing policy package of summer 2025 is attempting the same logic, at a scale appropriate to the urgency of the moment.

The OBBBA's $1.4 billion for the small UAS industrial base is not a research program. It is an industrial subsidy designed to bring the economics of domestic drone production into competitive range. The DOGE consolidation of Pentagon procurement, targeting at least 30,000 small drones under a single coordinating framework, is precisely the kind of large, predictable, volume purchase commitment that justifies capital investment in production infrastructure. The DoD procurement directive's acceleration of purchase timelines closes the remaining gap: not just volume, but velocity. The manufacturers who have spent years running pro formas on drone production economics now have the inputs they need to make the numbers work. The capital is flowing because the policy has made the demand legible.

The manufacturers who have spent years running pro formas on drone production economics now have the inputs they need to make the numbers work. The capital is flowing because the policy has made the demand legible.

The question that cannot yet be answered with confidence is whether the commitment will hold. A component supplier deciding in 2026 to co-locate with a domestic drone manufacturer's facility is making a capital allocation decision predicated on demand that will persist through 2030 and beyond, through election cycles, continuing resolutions, budget negotiations, and geopolitical shifts that may alter the calculus in ways that are today unforeseeable. The CHIPS Act's semiconductor investments are already generating results: TSMC's Arizona fabrication plant is producing chips, and the engineers who trained alongside that investment are now credentialed American workers. The drone manufacturing equivalent of that journey has just begun. The policy is real. The money is appropriated. The manufacturers are building. The durability of all three conditions, over the years-long timeline that industrial build-out requires, is the variable on which the entire enterprise depends.

The Jobs That Anchor Communities

When Skydio opens a facility five times larger than its current manufacturing operations and hires more than 5,000 workers, 2,000-plus in direct manufacturing roles, 3,000-plus across the supply chain, those jobs do not arrive in a vacuum. They land in a community. They generate demand for housing that raises property values and funds school districts. They create customers for restaurants, retailers, childcare centers, and healthcare providers. They attract engineers who bring families who join civic organizations, coach youth sports teams, and run for school board seats. The tax base expands. The local economy tilts, perceptibly and sustainably, toward higher-wage, higher-skill employment. A drone manufacturing cluster is not merely a line item in a defense appropriations bill. It is an economic development event with multiplier effects that reverberate through a region for decades.

This is the lesson that the semiconductor industry has been re-teaching, at considerable cost and with genuine urgency, since the passage of the CHIPS Act. The fabs are not merely production facilities. They are anchors for technical workforce ecosystems: community college programs redesigned around semiconductor process technology, apprenticeship pipelines aligned to clean-room manufacturing requirements, university research programs oriented toward the materials science and electrical engineering disciplines that sustain a semiconductor industry. The manufacturing investment created the demand for the workforce pipeline. The workforce pipeline made the manufacturing investment durable. The drone manufacturing build-out now underway must learn that lesson faster and apply it more intentionally than the semiconductor industry did, because the workforce constraint is already visible on the horizon.

The aerospace and defense manufacturing sector is already facing a 10 percent shortage of certificated mechanics and sustained engineering hiring challenges, pressures documented by the Aerospace Industries Association (AIA) and corroborated by workforce analyses from the Aviation Technician Education Council (ATEC) and Oliver Wyman. Drone manufacturing will draw from the same constrained talent pool. Every drone manufacturing facility needs engineers to design the systems. It needs technicians to assemble the airframes and calibrate the sensors. It needs quality control specialists to certify the components to the tolerances that defense and enterprise customers require. It needs supply chain managers capable of sourcing domestically through a supplier network that is still being built. Those positions are only viable if there exists, in or near the communities where those facilities open, a trained workforce prepared to fill them.

The manufacturing imperative and the talent imperative are not parallel challenges operating in separate institutional lanes. They are the same challenge, viewed from two different vantage points: the factory floor and the classroom. The communities that recognize this connection, that align their workforce development institutions, their community college programs, their apprenticeship systems, and their economic development strategies with the manufacturing investment that is now flowing into the drone industry, will be the economic winners of the uncrewed aviation era. The communities that treat manufacturing investment and workforce development as separate problems, to be addressed by separate agencies on separate timelines, will watch those opportunities settle somewhere more strategically prepared.

Part II turns from diagnosis to prescription.

Looking Ahead — Chapter Five

Part II turns from diagnosis to prescription. Chapter 5 examines the education, training, and certification infrastructure that will determine whether the workforce this industry urgently needs actually gets built, and by whom, and how fast.

Notes and Sources

1. White House. Executive Order: "Unleashing American Drone Dominance." June 6, 2025. Directs FAA to issue proposed BVLOS rule within 30 days,

prioritizes U.S.-made drones for federal agencies, and directs promotion of American-manufactured UAS exports.

2. One Big Beautiful Bill Act (OBBBA). Signed into law July 4, 2025. Provisions include $1.4 billion for small UAS industrial base scaling; $2.1 billion for medium unmanned surface vessel development; $188 million for maritime robotic autonomous systems R&D; $250 million for digital autonomy test environment.

3. DRONELIFE. "Pentagon DOGE Drone Program Targets 30,000 Small Drones Under Single Coordinating Framework." November 2025. Includes analysis of Replicator program outcomes and structural procurement reform efforts.

4. DRONELIFE. "Drone Investment 2025 Soars: Hundreds of Millions in Fresh Capital Tied to Legislative Signal." July 2025. See also: InnovateEnergy. "How New U.S. Laws Are Unleashing a Wave of Investment in American Drone Manufacturing." August 2025.

5. Bolt Flight. "2025 U.S. Drone Manufacturers: A Comprehensive Industry Survey." May 2025. Documents 60+ active U.S. drone manufacturers spanning defense, enterprise, and commercial segments.

6. Skydio, Inc. SkyForge Program Announcement. April 2026. Details $3.5 billion manufacturing expansion, 5,000+ jobs, $1 billion+ directed to domestic suppliers, Series F $110 million funding at $4.4 billion valuation. Includes delivery milestone of 60,000+ drones to 3,800+ customers including all U.S. military branches and 29 allied nations.

7. Defense Innovation Unit (DIU). Blue UAS Cleared List. August 2025. Lists 23 cleared drone platforms and 14 UAS framework components including Skydio, AeroVironment, Teal, Inspired Flight, Freefly, Teledyne FLIR, ModalAI, Shield AI, Parrot, Wingtra, and others.

8. International Energy Agency (IEA). Rare Earth Elements: Supply Chains, Markets and Energy Transition. 2024. Confirms China's 91% share of global rare earth refining output and 94% share of sintered permanent magnet production as of 2024. Available at iea.org. See also: U.S. Geological Survey, Mineral Commodity Summaries 2025, January 2025, confirming China's dominant position in rare earth processing. For the 99% drone battery cell figure: The Next Web. "US Banned DJI but China Controls 90% of Rare Earth Magnets and 99% of Drone Batteries Needed to Replace It." May 3, 2026. Available at thenextweb.com/news/us-drone-dji-ban-supply-chain.

9. American Security Drone Act (ASDA). Enacted December 22, 2023, as part of National Defense Authorization Act FY2024 (Public Law 118-31). Prohibits use of federal funds to procure or operate drones from covered foreign entities.

10. Federal Acquisition Regulation (FAR) Interim Rule. Effective November 12, 2024. Prohibits DoD, GSA, and NASA from procuring UAS from ASDA-covered foreign entities in contracts awarded on or after that date.

11. Federal Communications Commission (FCC). Covered List Expansion. December 2025. Expanded list to include DJI and broader foreign-manufactured UAS and critical UAS components, signaling regulatory direction toward component-level security requirements.

12. Dronesgator. "American-Made Drones: U.S. Manufacturers & NDAA Compliance Guide." Updated March 2026. Comprehensive reference on NDAA-compliant domestic drone manufacturers, Blue UAS cleared platforms, and procurement guidance for federal buyers.

PART II: BUILDING THE WORKFORCE

Future Flight

How Talent, Capital and Manufacturing will Shape the Next Era of Uncrewed Aviation

About Part II

Part I established the landscape: a drone industry expanding at historic velocity, a geopolitical competition in which workforce depth is a strategic variable, a talent deficit that threatens competitiveness, and a manufacturing imperative that demands a skilled domestic labor base. Part II turns to the prescription. Across four chapters, it examines the specific strategies, institutions, and communities positioned to build the workforce the industry requires: the education and credentialing pipeline, the untapped talent, the broader demographic opportunity the industry has yet to capture, and the employer strategies that will retain what the pipeline manages to produce.

Building the Pipeline: Education, Training, and Certification

"We have the students. We have the employers. We have the technology. What we are missing is the architecture that connects them, and architecture is a choice, not an accident."

— *Sharon Rossmark*

The Dean's Dilemma

On a Wednesday morning in early spring, Maria Delgado, the aviation program chair at a mid-sized community college in the Southwest, walked across her campus to the maintenance hangar where the school had just taken delivery of six new training drones. The machines were sleek, capable, and represented a meaningful investment, the kind of commitment her institution rarely made on the basis of a job market projection alone. Delgado had spent eighteen months making the case for them: compiling labor-market reports, presenting enrollment projections to a skeptical budget committee, writing grant applications that required her to translate the language of an industry barely a decade old into the familiar vocabulary of vocational education planning. She had won the argument. The drones were here.

What was not yet here was the curriculum to teach them with, approved and articulated in a form that employers would recognize. The faculty member she had recruited to teach the program's flight operations component was working with provisional adjunct status while the hiring committee completed its review. The articulation agreement with the regional four-year university that would allow her graduates to transfer seamlessly into a bachelor's program in aviation technology was still in draft, awaiting review by a curriculum committee that met every three months. The Federal Aviation Administration's (FAA) unmanned aircraft systems (UAS) Collegiate Training Initiative application, which would give the program formal federal recognition and open the door to the quarterly best-practice network with other CTI-designated schools, was complete, signed, and sitting in the submission queue.

Delgado was doing everything right. The frustration was not negligence, or lack of vision, or institutional indifference. The frustration was structural. The institutions most naturally suited to build the UAS workforce pipeline, community colleges, technical institutes, vocational programs, are, by design and by funding model, built for deliberate, committee-driven processes that prioritize stability, transferability, and broad accessibility. They are not built for the speed at which the commercial drone industry has expanded, and they are not resourced at the level that speed demands.

The hunger is real. The effort, at institutions like Delgado's, is genuine and impressive. The gap between institutional ambition and institutional capacity is not a gap of will. It is a gap of resources, timelines, and coordination infrastructure. And that gap, replicated across hundreds of programs at institutions across the country doing exactly what Delgado is doing, is where workforce shortages are born. Not in any single failure. In the accumulated friction of systems that were not designed to move together.

Which raises the animating question of this chapter: given the tools now available, a growing federal program, a credentialing ecosystem that is beginning to mature, an underutilized military pipeline, and instructive international models, what does a UAS workforce pipeline that actually works look like? And how far is the United States from having built one?

The Credential That Built an Industry, and Its Ceiling

It is worth pausing to appreciate what Part 107 accomplished, because the story of the UAS workforce cannot be told honestly without recognizing that the foundational commercial drone credential is, by almost any measure, an extraordinary policy achievement. When the FAA launched its Small Unmanned Aircraft Systems rule in August 2016, it did something that federal aviation regulation rarely does: it created a genuinely accessible on-ramp into a new aviation profession. For a $175 exam fee, any American sixteen years of age or older, no flight hours, no medical certificate, no aviation background required, could take a sixty-question aeronautical knowledge test and emerge as a certified commercial drone operator, legally permitted to fly for hire. Preparation can take many forms, from a semester-long community college course or a university program to any of the hundreds of commercial exam prep providers operating nationwide, making the credential accessible across virtually every learning style and schedule.

The numbers that followed were staggering. By the end of 2025, the FAA recorded 492,311 active Part 107 Remote Pilot certificates, a figure approaching the total number of non-student traditional pilots in the United States combined.3 That milestone reflects a wholesale reshaping of who participates in American aviation. Part 107 did not just create a credential. It created an industry's workforce, at speed, at scale, and with a cost barrier low enough that it did not function as a wall for the working adults most likely to pursue it.

Part 107 did not just create a credential. It created an industry's workforce, at speed, at scale, with a cost barrier low enough that it did not function as a wall.

The salary data attached to that credential reflects a genuinely attractive career proposition. Entry-level commercial drone pilots with basic Part 107 certification earn between $45,000 and $65,000 annually. Certified pilots with two to three years of real operational experience command $65,000 to $95,000. Senior certified operators reach $85,000 to $120,000, and the most experienced drone operators working in supervisory roles or leading drone operations companies earn $100,000 to $180,000 or more.8

But Part 107 was designed for a specific operational envelope, and the envelope is beginning to show its edges. The certification tests knowledge: aeronautical regulations, airspace classification, weather interpretation, emergency procedures. It requires no flight demonstration, no practical skills assessment, and no minimum hours in the air. A Part 107 certificate holder is a credentialed operator. They are not, by virtue of that credential alone, a skilled, experienced, or mission-ready professional. The distinction matters more with every passing quarter as the industry moves toward operational complexity that the original framework was not designed to address.

The BVLOS era, the period the industry is now entering in which beyond visual line of sight operations become the dominant commercial model, will require operators who function less like traditional pilots and more like air traffic controllers: managing fleets of aircraft simultaneously across distributed geographic areas, integrating data streams from multiple sensors, and making real-time decisions in airspace shared with crewed aviation. The Part 107 knowledge test was not designed to assess readiness for that environment. As the FAA's forthcoming Part 108 framework moves toward finalization, the gap between what Part 107 certifies and what advanced commercial operations require will become impossible to ignore.

Perhaps the most telling structural gap is the one that goes almost entirely undiscussed in public policy conversations about drone workforce development: the complete absence of any federally standardized credential for drone maintenance technicians. The FAA's Airframe and Powerplant (A&P) certificate is one of the most respected credentialing frameworks in American vocational education, a rigorous, nationally standardized, employer-recognized mark of competency that has defined the traditional aircraft maintenance workforce for generations. No equivalent exists for UAS. An employer hiring a drone technician today has no common standard against which to assess a candidate's knowledge and skills. No federal credential. No portable industry mark. No standardized wage structure tied to demonstrated competency. This is not a minor administrative gap. It is a systemic credentialing failure that makes UAS technical careers harder to value, harder to price, and harder to attract the next generation of workers into, at precisely the moment the industry needs maintenance technicians most.

The Federal Framework and Its Frontier

When Congress passed the FAA Reauthorization Act of 2018, Section 631 included a mandate with genuine potential: the FAA was directed to establish a Collegiate Training Initiative for unmanned aircraft systems, creating a formal federal framework for recognizing post-secondary institutions that committed to building UAS degree programs aligned with industry needs. The provision acknowledged what the industry had understood for some time, that the UAS workforce pipeline would not assemble itself through market forces alone, and that the federal government had both the authority and the obligation to create the architecture that would let educational institutions connect meaningfully to an industry transforming faster than curriculum committees could track.4

The program launched on April 30, 2020, with an initial cohort of twenty-six schools. The timing was not ideal, as the COVID-19 pandemic had shuttered campuses nationwide, but the institutions that joined in that first cohort did so with genuine commitment, and the program's growth over the five years that followed has been substantial. By January 2025, the FAA UAS-CTI roster had expanded to more than 145 institutions, a listing that spans nearly every region of the country and nearly every sector of post-secondary education.1 The thirty-eight pages of approved schools read like a deliberate survey of American higher education's geographic and institutional breadth: Aims Community College in Windsor, Colorado, and Clemson University in South Carolina; Carroll University in Waukesha, Wisconsin, and Central Michigan University in Mount Pleasant; technical schools in rural states and research universities in technology corridors.2

The program's architecture reflects careful thinking about how federal frameworks can support institutional capacity without imposing uniformity. To qualify for CTI recognition, an institution must be an accredited, not-for-profit post-secondary school, recognized by a federally recognized accrediting agency, and must offer a bachelor's or associate's degree in UAS, or a degree program with a UAS minor, concentration, or certificate. Public two-year colleges that join the program receive an additional designation as members of the Consortium for Small UAS Technology Training, gaining access to quarterly meetings with FAA representatives and a peer network through which curriculum developers at different institutions can share what is working. The program's explicit mandate, connecting schools with the FAA, industry partners, local governments, law enforcement agencies, and regional economic development entities to address labor force needs, acknowledges that pipeline building is inherently multi-institutional work.4

One hundred forty-five schools is impressive growth from twenty-six. It is still a fraction of the community college system that could and should be building UAS programs.

One hundred forty-five schools is impressive growth from twenty-six. It is still a fraction of the community college system that could and should be building UAS programs. The United States has more than 900 community colleges. The institutions that have joined the CTI program are, almost by definition, the ones that already had the institutional champions, the administrative energy, and the early equipment access to navigate the application process. The schools that haven't joined yet are often the ones in the communities where the need is greatest: rural agricultural regions where precision agriculture drone adoption is accelerating without a trained local workforce to support it; post-industrial communities where drone manufacturing and logistics could anchor economic revitalization if the labor supply existed to attract employers.

The curriculum challenge is equally significant. CTI recognition is a signal of institutional commitment. It is not, at present, a quality assurance mark that communicates to employers precisely what a graduate from a given institution has learned and demonstrated. The depth, rigor, and employer alignment of UAS programs at CTI-recognized institutions varies enormously, a function of differences in budget, faculty expertise, equipment access, and the maturity of local employer partnerships. A graduate from a program with active industry advisory boards, current equipment fleets, and structured internship pipelines is a materially different candidate than a graduate from a program that checked the CTI boxes but hasn't yet had the resources to build what sits underneath them. Until the CTI framework develops more robust curriculum standards and outcome metrics, the designation will function as a necessary but insufficient marker of pipeline quality.

The geographic alignment problem deserves particular attention. The UAS job market is not uniformly distributed, and it does not map neatly onto the existing distribution of CTI programs. Precision agriculture drone operations are concentrated in the Midwest and the South. Infrastructure inspection work follows the geography of American infrastructure, pipelines, transmission lines, bridges, rail corridors, across both rural and urban environments that may or may not have a nearby CTI institution. Public safety drone programs are proliferating in law enforcement agencies and fire departments serving communities of every size. Building a CTI network that intentionally closes these geographic gaps requires coordination that goes beyond the program's current voluntary, application-driven structure. It requires deliberate outreach to regions not yet reached by the CTI network, targeted funding for capacity building at institutions that need resources to apply rather than simply willingness to do so, and a data infrastructure that tracks where graduates are going and whether they are being hired by the employers that the industry needs them to reach.

From Semester to Credential

In February 2026, the Unmanned Safety Institute (USI) launched UAS Maintenance Technician programs at three community colleges: Central Arizona College, Horry-Georgetown Technical College in Conway, South Carolina, and West Georgia Technical College in Waco, Georgia. The launch received modest coverage in the UAS trade press. It deserved considerably more, because what USI and its partner institutions had built was not simply a new training program. It was a working prototype of the workforce pipeline model the industry needs at scale.5

The format was twelve weeks. Not four years. Not two. Twelve weeks of structured, intensive, hands-on training, covering electrical hardware diagnosis and repair, flight controller systems, electronic speed controller troubleshooting, autopilot data interpretation, composite airframe familiarity, and 3D-printed component maintenance, aligned throughout to ASTM International F3600 standards — the American Society for Testing and Materials standard that defines competency requirements for UAS maintenance technicians and is the most rigorous technical framework the industry has produced for this occupational category.7 Graduates earn the USI UAS Maintenance Technician Designation, a credential that carries meaning in the employer community precisely because it is built on a recognized, publicly available technical standard rather than a proprietary internal curriculum that only the issuing institution can explain or verify.

The program at West Georgia Technical College carried additional significance. When it launched in September 2025, it was the institution's first drone program of any kind, and it was described by its administrators as the first UAS maintenance technician program in the state of Georgia.6 Georgia had emerged as one of the more active states in agricultural drone adoption; its agricultural economy, and the precision agriculture technology companies serving it, had created genuine employer demand for maintenance technicians that the state's education system had not yet addressed. The USI program didn't wait for a four-year institution to build a comprehensive aviation technology department. It worked with an existing technical college, an existing workforce development infrastructure, and a clear technical standard, and it produced a twelve-week pathway from enrollment to employable credential.

The model's broader implications extend well beyond maintenance technician preparation. The accelerated, employer-aligned, standard-anchored program design is a template applicable across the full occupational range of the UAS industry. Fleet operations managers, remote sensor data analysts, UAS regulatory compliance specialists, airspace coordination professionals: most of the jobs the industry will need to fill in the next three to five years do not require four-year degrees. They require clearly defined competency standards, deliberately constructed curricula, delivery mechanisms that work for working adults, and credentials that employers can trust. The four-year university is not the wrong answer for every UAS career trajectory. It is the wrong answer for most of the hiring volume the industry needs to achieve in the near term, in the communities where those jobs will actually exist, for the candidates who need pathways that are accessible in months rather than years.

The USI model also addresses the credentialing gap that has made UAS maintenance careers difficult to value and hard to price. When there is no common standard, there is no common wage expectation, no reliable way for an employer to distinguish a well-trained candidate from an undertrained one, and no portable credential that moves with a worker across employers and states. The ASTM F3600 framework begins to solve that problem, not completely, and not for every occupational category, but in a meaningful and replicable way. Scaling the model from three community colleges to thirty, and from maintenance technicians to a broader range of UAS occupational categories, is within the reach of coordinated industry-education-government effort. Whether that effort materializes is, at this moment, an open question.

Warriors to Operators

Every year, tens of thousands of American military personnel separate from service carrying skills that the commercial drone industry urgently needs and consistently undervalues. The gap between the talent available in the military UAS community and the talent actually recruited into commercial drone roles is one of the largest and most consequential inefficiencies in the workforce development landscape, and it is almost entirely a coordination failure rather than a capability gap.

Army Military Occupational Specialty (MOS) 15W, Unmanned Aerial Vehicle (UAV) Operator, is, by the assessment of military transition specialists and commercial drone employers alike, among the most directly translatable military occupational specialties to commercial UAS work in the entire U.S. armed forces. The Military Transition Toolkit's January 2026 guide to the 15W-to-commercial career pathway describes the transfer as essentially direct.8 A 15W operator leaving active duty carries with them the demonstrated ability to operate complex uncrewed systems in dynamic, high-stakes environments; manage payload sensors and interpret their outputs; navigate and comply with airspace regulations of significant complexity; maintain detailed flight logs and equipment maintenance records; assess weather conditions for flight planning; communicate with air traffic control; make safety-critical decisions under pressure with high-value equipment; and bear full operational responsibility for missions where failure has real consequences. These are not adjacent skills to the core competencies of a professional UAS operator. They are those competencies, developed at government expense, validated in operational environments, and ready to deploy the day after a service member's separation paperwork clears.

"Military drone operators bring maturity, discipline, and regulatory understanding that civilian self-taught pilots cannot match." — Military Transition Toolkit, 2026

The credential barrier into commercial work is minimal. Part 107 certification is straightforward for veterans with active UAS operational experience; the aeronautical knowledge the test assesses is knowledge they have been applying in the field. The salary ceiling on the other side of that credential, $100,000 to $180,000 or more for experienced operators in senior commercial or supervisory roles, is competitive with civilian career options that typically require years of additional education and credentialing to achieve. The commercial drone industry is actively seeking exactly the combination of technical competency, operational discipline, and regulatory awareness that military UAS experience produces. And it is largely failing to find it, not because the talent does not exist, but because the industry and the military separation system are not communicating effectively enough to connect them.8

The 15W MOS is the most widely recognized gateway, but it is far from the only relevant military pipeline. Army MOS 15E, uncrewed aircraft systems repairer, produces technical maintenance professionals whose skills map directly to the maintenance technician gap the USI programs are beginning to address. MOS 153D covers UAS pilot warrant officers, a population with even deeper operational experience and leadership competency. The Navy and Air Force have their own equivalents: a broad and largely uncatalogued population of technically trained drone professionals whose skills are inadequately mapped to civilian occupational frameworks, inadequately marketed to civilian employers, and inadequately supported through the transition process.9 The Military Transition Toolkit's 2026 assessment describes the situation frankly: the commercial drone industry is leaving one of its most immediately available talent pipelines almost entirely on the table.

What capturing this pipeline requires is not complicated in concept, though it demands sustained organizational effort. Military UAS operators need transition programs specifically designed for their career context, programs that go beyond resume translation and generic job search coaching to connect departing service members with commercial drone employers before separation, help them understand how their MOS maps to civilian job titles and salary bands, guide them through the Part 107 certification process, and provide the community and mentorship structures that ease the cultural transition from military aviation to a commercial industry whose vocabulary, hierarchy, and operational norms differ in significant ways from the ones they are leaving. The FAA CTI network, the USI credentialing framework, and the growing universe of employer-led training programs are the raw materials of this bridge. They are not yet assembled into anything resembling a functional system. Building that system, deliberately, with dedicated resources and genuine employer commitment, would represent one of the highest-return workforce investments the industry can make. The talent has already been trained, at public expense, and it is walking out the door of military bases every year in search of a clear path forward.

The Progressive Ladder

The United States did not invent commercial drone regulation, and it has not, in the area of progressive workforce credentialing, kept pace with the frameworks that peer nations have developed. The comparison is instructive, not as a rebuke, but as a mirror that shows clearly what the American system is missing and what it would take to build it.

Canada's 2025 regulatory expansion, effective November 4, created a new Level 1 Complex BVLOS certification framework that illustrates the principle of progressive, demonstrated-competency credentialing with particular clarity. Pilots seeking Complex BVLOS authorization must complete twenty hours of ground school and pass a demonstrated flight review, a practical assessment of operational competency, not merely a written test of regulatory knowledge. Entities conducting BVLOS operations must obtain a new Remotely Piloted Aircraft System (RPAS) Operator Certificate, with technical standards calibrated to the complexity and risk profile of the operations they intend to conduct.10 Transport Canada's framework was explicitly designed in two phases to avoid disrupting the peak summer flying season, a detail that reflects the practical sophistication of a regulatory body thinking carefully about how certification requirements interact with actual operational calendars.

The United Kingdom's 2025 Remote Pilot Certificate (RPC) framework takes the progressive credentialing model further. The Civil Aviation Authority (CAA) established four levels: RPC-L1 for basic visual line of sight (VLOS) competence, advancing through RPC-L2 and RPC-L3 for increasingly complex BVLOS operations in integrated airspace environments, with RPC-L4 as a future pathway for expert-level, highly complex BVLOS work, each requiring demonstrated flight hours at the preceding level before advancement is permitted.11 Training and assessment are delivered through a network of CAA-approved Recognised Assessment Entities, creating a quality-assured delivery infrastructure without requiring the regulator to deliver training directly. The result is a credential ladder that a pilot can climb from initial certification through commercial professional complex operations specialist, with each rung representing a demonstrated, assessed, and nationally standardized level of competency.

What the United States has, by comparison, is a single-step credential, Part 107, followed by a long, unstructured expanse of self-directed experience accumulation with no standardized milestones, no progressive credentialing ladder, and no national mechanism by which an employer can assess where on that expanse a given candidate actually stands. The BVLOS era will demand more. The Part 108 framework, when finalized, will create new operational authorization requirements. But authorization is not the same as a workforce credentialing system, and the experience of Canada and the UK suggests that the two need to be developed in tandem if the commercial industry is to have the qualified workforce it needs when the regulatory door opens.12

The architecture for an American progressive credentialing system is not absent. It is disaggregated. The FAA UAS-CTI network provides the institutional foundation: 145 accredited schools, connected to the federal agency, operating across the full geographic range of American higher education. The ASTM F3600 standard provides a technical framework on which employer-recognized maintenance credentials can be built. The USI twelve-week model provides a curriculum prototype that demonstrates how accelerated, standard-aligned training can produce portable credentials at community college scale. The military pipeline provides a ready and genuinely skilled talent base. What is missing is the coordination that converts these elements from parallel initiatives into a coherent national system: clear entry points that a prospective UAS worker can identify and navigate; progressive milestones that carry meaning across employers and states; portable credentials at every level of occupational complexity that allow workers to move through the industry and allow employers to assess what they are hiring; a national architecture that the industry, the government, and the education system have built together rather than in isolation from one another.

Building that architecture is the defining workforce development task of the next five years. The institutions exist. The talent exists. The frameworks exist. The coordination does not, yet. That is not a technology problem. It is a leadership problem. And it is one for which the tools are already in hand.

Looking Ahead — Chapter Six

Women hold only approximately 9 percent of active Part 107 Remote Pilot certificates, even as that number is growing at a 19.3 percent year-over-year rate that suggests a momentum the industry has barely begun to cultivate. Chapter 6 asks not merely why this gap persists, but what happens, to the industry's capacity, its ability to recruit across the full range of communities it needs to reach, if it does.

Notes and Sources

1. Federal Aviation Administration. UAS Collegiate Training Initiative (UAS-CTI) 5-Year Review. FAA, 2025. Covers program growth from 26 inaugural schools (April 2020) to 145+ by January 2025, program architecture, and the Consortium for Small UAS Technology Training designation for public two-year colleges.

2. Federal Aviation Administration. FAA Approved UAS Collegiate Training Initiative Schools. Official listing, May 2025. 38-page document spanning all U.S. regions; institutions cited include Aims Community College (Windsor, CO), Clemson University (SC), Carroll University (WI), and Central Michigan University (MI).

3. Federal Aviation Administration. Civil Airmen Statistics — Active Remote Pilot Certificates (Part 107). FAA, December 31, 2025. Reports 492,311 active Part 107 certificates.

4. Federal Aviation Administration. UAS Collegiate Training Initiative Program Overview. See also: Section 631, FAA Reauthorization Act of 2018, Pub. L. 115-254. Legislative mandate for FAA to establish the UAS-CTI program.

5. Unmanned Safety Institute. USI Launches UAS Maintenance Technician Programs at Three Community Colleges. USI press release, February 2026. Covers 12-week program format at Central Arizona College, Horry-Georgetown Technical College (SC), and West Georgia Technical College (GA); alignment to ASTM F3600 standard; USI UAS Maintenance Technician Designation.

6. West Georgia Technical College / Unmanned Safety Institute. West Georgia Technical College Launches First Drone Program in Partnership with USI. Program announcement, September 2025. Described as the first UAS maintenance technician program of its kind in Georgia.

7. ASTM International. F3600 Standard Specification for UAS Maintenance Technicians. ASTM International. The technical framework underlying the USI maintenance technician credential and employer-recognized industry standard for UAS maintenance competency.

8. Military Transition Toolkit. Army MOS 15W UAV Operator to Commercial Drone Pilot Career Pathway Guide. January 19, 2026. Source for skills transfer

analysis, salary ranges ($45,000–$180,000+), Part 107 transition pathway, and the quoted assessment of military operator competencies relative to civilian self-taught pilots.

9. Military.com. Military Drone Pilots: Skills Transfer to Civilian UAS Careers. April 2, 2026. Analysis of military UAS occupational specialties (15W, 15E, 153D, Navy and Air Force equivalents) and structural barriers to civilian pipeline capture.

10. Transport Canada. Summary of Changes to Drone Regulations — Advanced BVLOS and Complex Operations. Transport Canada, 2025. Covers Level 1 Complex BVLOS certification framework effective November 4, 2025; 20-hour ground school requirement; flight review; new RPAS Operator Certificate (RPOC); two-phase implementation timeline.

11. UK Civil Aviation Authority. Remote Pilot Certificate (RPC) Framework: Levels 1–4. CAA, 2025. Covers the four-level progressive RPC structure (RPC-L1 through RPC-L4), demonstrated flight hour requirements at each level, delivery through CAA-approved Recognised Assessment Entities (RAE-PCs), and the BVLOS progression pathway.

12. McDanolds, James. "Part 108 and the Workforce It Will Demand." Aviation International News (AIN). April 21, 2026. Analysis of BVLOS era workforce implications of Part 108 rulemaking, credentialing gaps relative to international frameworks, and the distinction between operational authorization and progressive workforce credentialing.

Half the Workforce

"An industry that recruits from half the talent pool and wonders why it can't find enough workers has diagnosed the symptom and ignored the cause."

— *Sharon Rossmark*

The Numbers Behind an Avoidable Shortage

The UAS industry faces a talent shortage it has, in part, manufactured for itself. Women represent approximately 50 percent of the working-age population. As of December 31, 2025, women hold 44,274 of the 492,311 active FAA Part 107 Remote Pilot certificates in the United States, roughly 9 percent.[1] In an industry projecting a shortage of hundreds of thousands of workers within years of Part 108 finalization, that gap is not a social statistic. It is a supply-chain problem. The industry cannot simultaneously claim a talent crisis and a rational explanation for drawing its certified commercial workforce from fewer than one in ten members of half the available labor pool. The gap is not immutable. Women's Part 107 certification grew 19.3 percent year-over-year in 2025, outpacing the overall market's 15.1 percent growth.[2] The talent is responding. The question is whether the industry will meet it.

What the Number Tells Us, and What It Hides

Begin with what is genuinely worth celebrating. That 19.3 percent year-over-year growth represents a gain of 7,161 women from 37,113 certificates at the close of 2024 to 44,274 at the close of 2025. The headline embedded in that comparison is significant and too rarely stated plainly: women are not simply growing in parallel with the industry. They are growing faster than it. They are gaining share. In an industry accustomed to framing gender participation as a lagging indicator, something that improves slowly and as a byproduct of broader growth, the 2025 data tell a different and more urgent story.

The geographic distribution of that growth adds a layer of analytical precision that the aggregate number alone obscures. The ten states with the highest percentage growth in women's Part 107 certification in 2025, among states with at least 100 women pilots in the prior year, were South Dakota, Indiana, Tennessee, Wisconsin, Alabama, Louisiana, Colorado, Florida, Washington, and California, with South Dakota leading at 27.3 percent growth and Indiana and Tennessee following at 26.6 and 25.7 percent respectively.[3] These are not the coastal technology corridors that dominate most conversations about the drone industry's future. They are agricultural states, Southern states, Midwest states, precisely the regions where drone adoption in precision agriculture, infrastructure inspection, and rural public safety is accelerating most rapidly. The pattern is not coincidental. Where drones are most immediately and concretely economically relevant, women are finding pathways in. The technology's utility is creating its own recruitment mechanism, and women are responding to it.

Table 6.1 — Top 10 States by Women's Part 107 Certification Growth, 2025

State	% Growth (Women, 2024–2025)	Primary Drone Adoption Drivers
South Dakota	+27.3%	Precision agriculture, infrastructure
Indiana	+26.6%	Ag tech, logistics, manufacturing inspection
Tennessee	+25.7%	Public safety, infrastructure, logistics
Wisconsin	+25.1%	Agricultural mapping, forestry
Alabama	+25.0%	Infrastructure, public safety
Louisiana	+24.6%	Energy sector inspection, coastal monitoring
Colorado	+20.2%	Public safety, environmental monitoring
Florida	+19.9%	Infrastructure, real estate, coastal
Washington	+19.0%	Forestry, agriculture, tech sector
California	+18.0%	Broad commercial adoption, tech integration

Source: FAA Civil Airmen Statistics, December 31, 2025. States with at least 100 women pilots in prior year. Primary drone adoption drivers reflect the author's assessment based on regional industry data and published drone market analyses; this column is not derived from FAA certification data.

The average age of entry is equally instructive, and it complicates a narrative the industry sometimes tells itself about pipeline problems. The average American woman commercial drone pilot is approximately 39 years old, with the largest concentration of participation in the 25-to-44 age cohort. This is not, in the main, a story about young women choosing drone careers fresh out of high school or aviation programs. It is a story about mid-career professional women, environmental consultants, public safety professionals, surveyors, journalists, real estate agents, educators, healthcare workers, discovering that drone technology offers a powerful and economically compelling extension of expertise they already possess. They are not reinventing themselves. They are adding a tool that transforms what they can already do. The pathway into UAS, for a very large share of the women entering it, runs through the work they are already doing.

The average American woman commercial drone pilot is approximately 39 years old. This is not a story about young women choosing drone careers fresh out of high school. It is a story about mid-career professionals discovering that drone technology transforms what they can already do.

And then there is what the 9 percent figure conceals. Part 107 certification is an entry point, not an outcome. Having a certificate establishes a legal right to fly commercially; it does not establish career trajectory, earning parity, leadership access, or retention. The Oliver Wyman/International Aerospace Womens Association (IAWA) "Lift Off to Leadership" study, released in 2025 as its second edition and drawing on a survey of 250 industry leaders globally, maps the landscape of what follows initial entry with uncomfortable specificity.[4] Women in aerospace and aviation were half as likely as men to rate gender equality in their organization as high or very high. Twice as many women as men believed their company placed little or no priority on increasing women's representation in leadership. Nearly half of the women surveyed had considered leaving the industry at some point in their careers, not because of the work, but because of the environment in which they were being asked to do it. Forty percent cited documented career advancement barriers as a growing and specific factor in their career trajectory. Thirty-eight percent reported being called "too aggressive" during their careers, compared to 15 percent of men, a data point that does not require interpretation.

The Collegiate Aviation Review International's 2025 comprehensive analysis of women's participation across all pilot certification levels, a mixed-methods study drawing on 31 scholarly sources and a full decade of FAA data, identifies five primary structural barriers with the kind of methodological precision that makes them difficult to dismiss as anecdote: recruitment pipeline limitations, financial constraints, masculine organizational culture, mentorship deficits, and work-life integration challenges.[5] Sixty-two percent of women in aviation reported experiencing harassment. Fifty-one percent faced retaliation when they reported it. At the ATP certification level, the highest credential in traditional aviation, achieving gender parity would require, at current advancement trends, more than 340 years. The UAS industry does not have 340 years. It barely has five before the talent shortage it is already experiencing becomes a talent crisis that no amount of automation can fully substitute for. The leaky pipeline is not a metaphor. It is a mechanism, and it operates at every level above entry.

A New Industry, an Old Pattern, and a Real Opportunity

The argument for urgency in the UAS industry specifically rests on a distinction that is real, meaningful, and in danger of being squandered. The UAS industry is new. Traditional aviation's gender gaps were not accidents; they were built, systematically and over a century of institutional practice. Military flight training explicitly excluded women until 1976, when the Air Force admitted women into its pilot training program for the first time.[10] The cultural mythology of the aviator, the leather jacket, the laconic confidence, the aggressive physical courage, was constructed as an aggressively male archetype and embedded in the institutions, the training culture, the informal networks, and the promotional materials of an industry that then expressed puzzlement at why women made up such a small share of it. These are not ancient artifacts of a forgotten era. They are the sediment that still shapes the culture of aviation organizations today, the inherited assumptions that persist in leadership styles, in mentoring relationships that form along lines of similarity, and in the unexamined standards that define who looks like a pilot, who sounds like an expert, and who gets the benefit of the doubt in a technical room.

The UAS industry represents a new sector of aviation, one that did not inherit the structural exclusions that traditional aviation built over a century. It started differently, and that difference matters more than it is usually given credit for. Part 107 requires no flight hours accumulated over years of expensive instruction. It requires no instructor relationship, no access to a flight school, no investment in an aircraft or a club membership or a social network built through those channels. It requires knowledge and the discipline to prepare for an exam that is accessible to anyone. The credential is democratic in a way that the Airline Transport Pilot (ATP) certificate is not and was never designed to be. The result, 19.3 percent year-over-year growth in women's certification rates, outpacing the overall market in 2025, is not mysterious. It reflects what happens when access is genuinely equal: qualified people pursue the opportunity.

When access is genuinely equal, qualified people pursue the opportunity. The 19.3 percent growth rate in women's Part 107 certifications is not a mystery. It is what equal access looks like in practice.

But the 9 percent ceiling, and it is, for the moment, effectively a ceiling, reveals that lowering the entry barrier is necessary and not sufficient. Women are entering the UAS industry at an accelerating rate. They are not advancing through it at a comparable rate. The cultural barriers documented in the Oliver Wyman/IAWA study do not have a sector exemption. They do not disappear because the airframe is uncrewed or because the company was founded in 2018 rather than 1968. Microaggressions travel across industry boundaries. Structural barriers to advancement operate in hiring committees that have never flown a traditional aircraft. The leadership representation gap is a function of organizational culture, not of an industry's age. An industry that wants to be different from its predecessors has to actively choose to be different, in its hiring practices, its leadership development investments, its organizational norms, and, critically, its financial support for the organizations that are building the pipelines it says it needs.

The opportunity is real and it is time-limited. The UAS industry is at the precise moment in its institutional development, before the culture calcifies, before the informal networks become entrenched, before the next generation of leadership is selected by the same homogeneous cohort that selected the current one, when deliberate choices about representation have their maximum leverage. Industries that miss this window do not simply fail to be equitable. They fail to be competitive. They narrow their own talent pool in a market environment where the talent pool is already the binding constraint on growth. The choice is not between equity and strategy. It is between recognizing that they are the same thing or learning that lesson expensively.

The Funding Is There

One of the most persistent and least defensible claims in the conversation about women's representation in the UAS workforce is that the funding to address it isn't available. It is available. What is in shorter supply is the organizational capacity to find it and the institutional knowledge to apply for it effectively. The FAA awarded $13.5 million in 22 Aviation Workforce Development Grants in fiscal year 2024.[6] The program's mandate is explicit and directly applicable: support recruitment of aircraft pilots and UAS operators through education, outreach, recruitment, and military career transition. The FAA Aircraft Pilots Workforce Development Grants Program specifically authorizes funding for community outreach, awareness programs, training access, and mentorship infrastructure, precisely the program types that produce measurable results in expanding the certified workforce. This is federal money, publicly authorized and specifically targeted for this purpose, and it is underutilized by the organizations doing the most effective work in this space.

The Department of Labor's Workforce Innovation and Opportunity Act (WIOA) funds add another substantial and largely untapped layer of accessible capital.[7] WIOA is specifically designed to support training and workforce development for workers entering or re-entering the labor market. Women transitioning into nontraditional technical careers, mid-career professionals retraining for emerging industries, and workers in rural communities accessing job markets that are expanding faster than local training infrastructure can serve them, all of these populations fit squarely within WIOA's mandate, and all of them are candidates for UAS workforce programs. The UAS industry has not yet fully developed the organizational literacy to engage with workforce development funding streams, to translate its talent needs into the language of federal program priorities, to frame its workforce development activities in terms that unlock investment rather than exclusively in the terms of commercial urgency that are more natural to a technology sector. That literacy is learnable, and it is worth learning.

The funding to expand the UAS talent pipeline is not an obstacle. It is an underutilized resource. The organizations that develop the capacity to navigate these funding streams will have access to a capital base that conference sponsorships and membership fees cannot match.

SBA programs for women business owners have historically provided a meaningful parallel pathway, one whose relevance to the UAS industry is still underappreciated. Women-owned drone service businesses are growing in precision agriculture, media production, infrastructure inspection, real estate, and public safety consulting. The entrepreneurs building these companies often lack access to the capital networks that male-dominated tech startups navigate more easily, not because capital is unavailable to them but because the connective tissue, the introductions, the warm referrals, the pattern-matching that venture networks rely on, has not historically been built with them in mind. The SBA's Women's Business Centers, which have historically provided exactly this kind of connective support, are facing proposed elimination in the FY2026 federal budget. Should Congress preserve them, they remain a meaningful resource. In their absence, private women-focused business accelerators and industry organizations will carry a greater share of this work.

The policy and funding landscape for expanding women's participation in the UAS workforce is navigable, though it is changing. FAA workforce development grants and WIOA funding remain active mechanisms for organizations building talent pipelines. NSF STEM education programs have historically been a significant resource, though deep proposed budget cuts and grant terminations beginning in 2025 have created uncertainty about their availability; organizations planning to rely on NSF funding should verify current program status before building program strategies around it. SBA entrepreneurial development programs face similar uncertainty. The employers, training providers, and industry organizations that develop the capacity to identify and navigate whichever funding streams remain available will have access to a capital base that sustains talent pipeline programming at a scale that conference sponsorships and membership fees alone cannot match. The gap is not primarily in the availability of funding. It is in the awareness of where to look, the capacity to apply, and the organizational will to treat workforce expansion as a strategic priority rather than an afterthought.

This Is a Business Decision

The competitive argument for women's participation in the UAS workforce does not require moral framing to be compelling, though the moral case is sound and should be stated plainly. The numbers are sufficient. The UAS industry faces a talent shortage of hundreds of thousands of workers, projected to intensify within years of Part 108 finalization. Women represent approximately 50 percent of the working-age population. The industry currently draws its certified commercial workforce from approximately 9 percent of that half. Whatever the mechanism, cultural barriers, visibility deficits, recruiting practices that default to established networks, organizational cultures that attract women and then fail to retain them, the arithmetic is stark. Every point of women's participation that the industry fails to capture is a constraint it has imposed on itself, a voluntary narrowing of the talent pool in a moment when the talent pool is the binding constraint on growth.

The industries where drone adoption is accelerating fastest are not industries where women are absent. They are industries where women are already working, already credentialed, already building careers and developing the domain expertise that makes a skilled drone operator substantially more valuable than a generalist remote pilot. The environmental field scientist who has spent fifteen years collecting data in difficult terrain does not need to be persuaded that drone mapping is transformative. She needs to be persuaded that the path to certification is accessible, that the companies in this space will take her seriously, and that the career on the other side of the Part 107 exam is worth the investment of her time and attention. The first condition, an accessible credential at a democratized entry cost, is largely met. The second and third conditions are precisely what the organizations doing this work are building, and they are building them faster than the industry is supporting them.

At age 39, the average entry point for women into commercial drone careers represents not a pipeline problem but a visibility and access problem of a specific and solvable character. These are not women who lack skills, credentials, or professional ambition. They are women in adjacent fields who have not yet seen a clear and credible signal that the UAS industry is looking for them, values what they bring, and will offer them a career rather than a credential. That signal, credible, consistent, delivered through visible role models, accessible programs, and organizations with demonstrated track records, is what changes the calculation. Building it is within the reach of any employer, training institution, or industry association that decides it is worth building.

The argument for acting is not complex. It does not require new research, new regulatory frameworks, or new categories of institutional infrastructure. The industry needs workers. Women are available, responsive, and growing their participation faster than the overall market. The organizations that know how to reach them exist and have demonstrated results. The federal funding to support those organizations is authorized and accessible. The credentialing pathway from community engagement to accredited degree now connects in ways it did not five years ago. What remains is the decision, by employers, by educational institutions, by investors and industry associations and the executives who set the tone for a sector's culture, to treat women's participation not as a nice-to-have that gets addressed after the more urgent problems are solved, but as a must-have that gets addressed first precisely because everything else depends on it.

The 9 percent is not a ceiling. It is a starting point. Every point of women's participation that the industry fails to capture is a constraint it has imposed on itself, a voluntary narrowing of the talent pool in a moment when the talent pool is the binding constraint on growth. The women in that room, the student with her fresh certificate, the veteran with her decade of operational experience, the environmental consultant who mapped a watershed with a drone last Tuesday and published the results this morning, are evidence of what is possible when the industry decides to look for what it says it needs. The question is not whether the talent exists. It does. The question is whether the industry's will to find it is as urgent as the need.

Looking Ahead — Chapter Seven

The regions with the largest concentrations of workers available to the industry, agricultural communities, post-industrial towns, rural corridors where drone adoption is accelerating fastest, are precisely the regions where the formal pipeline is thinnest. Chapter Seven examines what it takes to reach workers in these communities: the programs that produce measurable results, the community college pathways showing early gains, and what happens to an industry's capacity when it finally commits to going where the talent actually is.

Notes and Sources

1. Federal Aviation Administration. Civil Airmen Statistics, December 31, 2025. FAA Aeronautical Center, Oklahoma City, OK. Primary source for 2025 Part 107 certification statistics cited throughout this chapter, including total active certificates (492,311), women holding active certificates (44,274), year-over-year growth rates, and historical comparison data.

2. FAA Civil Airmen Statistics, December 31, 2025; Skyfarer Academy 2025 FAA Data Hub. Source for traditional pilot certificate data (100,704 women holding traditional pilot certificates, approximately 11.4% of total) and year-over-year remote pilot growth rate comparisons (overall: 15.1%; women: 19.3%).

3. Federal Aviation Administration. Civil Airmen Statistics, December 31, 2025 — State-Level Data. FAA Aeronautical Center. Source for state-by-state women's Part 107 certification growth data, including top-10 state rankings by percentage growth and the primary drone adoption context for each state.

4. Oliver Wyman / International Aerospace Womens Association (IAWA). "Lift Off to Leadership: Leadership Barriers for Women in Aerospace and Aviation." 2nd edition, 2025. Survey of 250 global industry leaders. Source for data on gender equality ratings, leadership representation, career departure consideration rates, career advancement barrier findings, and the "too aggressive" finding.

5. David Ison. "Women's Participation in U.S. Pilot Careers: A Mixed-Methods Analysis of Trends, Barriers, and Workplace Climate (2015–2024)." Collegiate Aviation Review International, Vol. 43, No. 2, September 9, 2025. Mixed-methods analysis drawing on 31 scholarly sources. Source for five-barrier framework,

harassment and retaliation statistics, ATP parity projection, and student pilot representation trend data.

6. Federal Aviation Administration. "Funding Opportunities — Aviation Workforce Development Grants." FAA.gov, updated January 8, 2026. Source for $13.5 million / 22 grants figure for fiscal year 2024 and program mandate description.

7. U.S. Department of Labor. Workforce Innovation and Opportunity Act (WIOA) grant programs. DOL.gov. Source for WIOA mandate, eligible populations, and applicability to UAS workforce training programs.

8. Aviation International News / James McDanolds, School of Uncrewed Technology, Sonoran Desert Institute. April 21, 2026. Source for broader UAS workforce shortage projections and commercial drone market growth context.

9. Federal Aviation Administration. "Civil Airmen Statistics: Historical Data." FAA.gov, accessed April 2026. Longitudinal source for women's pilot certificate data across certification levels and year-over-year trend analysis cited in comparison with UAS certification data.

10. National Women's History Museum. "Women Airforce Service Pilots (WASPs) of WWII." womenshistory.org. Notes that the U.S. Air Force accepted women into its pilot training program in 1976. Corroborated by: Smithsonian National Air and Space Museum. "Documenting Change: The Museum's Military Women Aviators Oral History Initiative." airandspace.si.edu. Documents the first class of USAF women pilots, UPT Class 77-08, who completed training in 1977 following the 1976 program opening.

From STEM to Flight: Reaching the Full Depth of the Talent Pool

"The talent has never been the problem. The pipeline has. And a pipeline that doesn't reach the communities where the jobs are growing isn't a pipeline at all: it's a bottleneck with a different name."

— Sharon Rossmark

The Student Who Saw the Number

Leonardo Fernandez was a senior at Southwest High School in San Antonio, Texas, when he first understood what a six-figure income meant. Not in the abstract way that students sometimes encounter the concept in a personal finance lesson, a number on a worksheet that feels as hypothetical as a lottery jackpot. Leonardo encountered it as a possibility, his possibility, attached to a credential he could earn before he graduated, in a program running down the hall. The program was a drone pilot training course supported by the Dee Howard Foundation, operating inside San Antonio's Southwest Independent School District. The credential was the FAA Part 107 Remote Pilot Certificate. And the number was not hypothetical at all.

Southwest ISD is the kind of school district that does not appear prominently in the marketing materials of the aviation industry. The majority of its students are Hispanic, many from low-income families, and the economic geography of the South Side of San Antonio is not one that has historically produced commercial pilots or aerospace engineers in large numbers. That is not a statement about the students. It is a statement about the pipelines, or rather, the absence of them. For most of the young people who grow up in San Antonio's Southwest ISD, the aviation industry has existed at approximately the same distance as the moon: visible on clear nights, theoretically reachable, practically inaccessible.

The Dee Howard Foundation changed that calculus, at least for the students in more than twenty high schools across the district who now have access to drone pilot training programs focused on basic STEM skills, flight simulation, and direct preparation for the FAA Part 107 commercial pilot certification exam.[1] Leonardo Fernandez enrolled because the program offered something he had genuinely never seen made available to someone who looked like him, who came from where he came from: a professional credential with a credible economic payoff. A six-figure income was, as he described it, "unheard of" for his family. The drone program made it hearable.

Rodolfo Urby, the aviation director at Southwest High School, speaks about what programs like this one mean in language that is careful, measured, and precisely right. "I think for kids," Urby has said, "the biggest thing for families here, especially in our district, is going to be opportunity, exposing them to all sorts of different things they may not have been exposed to."[2] What Urby is describing is not a job training program. It is a horizon-expansion program. It is what happens when an institution decides that the young people in its care deserve to see the full landscape of what their lives could become, including the parts of that landscape that nobody in their families or neighborhoods has yet mapped.

That is what drone education, at its best, does for communities the aviation industry has not historically reached. It does not merely offer a credential. It offers a proof of concept: that the industry is reachable, that the credential is attainable, that the career is real and the income is real and the path from where you are standing right now to where you want to go is shorter and less expensive than you had been led to believe. Leonardo Fernandez enrolled in a drone program. He saw a number. He understood, perhaps for the first time, that the number had his name on it.

This chapter makes a practical argument, rooted in that moment in a San Antonio classroom. The regions the UAS industry has most systematically failed to reach, rural agricultural corridors, post-industrial towns, lower-income urban neighborhoods, the K–12 students who are growing up right now and will form the industry's workforce in five and ten and fifteen years, are precisely the regions the industry most urgently needs. The argument is a competitive necessity argument. The talent is there. The pathways are not. That is not a permanent condition. It is a solvable problem, and the evidence that it is being solved, unevenly and incompletely but with genuine momentum, is already accumulating in classrooms and community centers and technical colleges across the country.

Where the Jobs Are, Where the People Are, and Why They Aren't Connected

There is a persistent misconception embedded in the way the UAS industry talks about its workforce gap. The conversation tends to center on coastal innovation hubs, the corridors around Silicon Valley, the research universities of the Northeast, the defense contractors clustered around Washington, D.C., as if the drone economy were primarily a coastal phenomenon requiring a coastal workforce. It is not. The economic geography of uncrewed aviation is far more distributed, and far more rural, than the industry's marketing materials tend to suggest. Precision agriculture drone operations are centered in Iowa, Kansas, Nebraska, Georgia, and the Mississippi Delta. Infrastructure inspection follows America's pipelines, power lines, and bridges into rural and semi-rural corridors that run through some of the least economically advantaged counties in the nation. Public safety drone programs are expanding in cities of every size, including cities where large numbers of young residents have never heard of the FAA Part 107 certificate.

The drone economy is dispersed. The talent pipeline has been built for somewhere else.

The structural consequences of that mismatch are not abstract. The U.S. Government Accountability Office's July 2025 report on STEM education for rural populations, GAO-25-107371, makes the challenge concrete in ways that are difficult to set aside.[4] The report documents a rural Alabama school district that entered the 2023–24 academic year with no certified math teachers available for three hundred students in grades seven through twelve. Not a shortage of math teachers. No certified math teachers. If a district cannot reliably staff a certified math classroom, its chances of staffing a drone pilot education program, which the FAA requires to be led by FAA-certified teachers, are correspondingly remote.

The GAO report identifies the cost and logistics of traveling long distances as a compounding barrier in rural communities. A field trip that a student in a well-funded suburban district takes for granted, a visit to an aerospace facility, a university research lab, an industry career fair, requires, in many rural communities, a level of transportation, funding, and administrative coordination that is simply not available. The hands-on experiences that turn an abstract career possibility into a concrete aspiration are more difficult to access precisely where they are most needed: in the communities where drone-based careers in precision agriculture, rural infrastructure inspection, and emergency response are most directly economically relevant.

In urban communities, the barriers are different in character but equally real in effect. The story of Frankford High School's Aviation Academy in Northeast Philadelphia is instructive. Josh Bergerson, the aviation and earth science teacher who leads the program, reported that student interest doubled when drones were introduced into the curriculum. Sixty students were already committed for the following year's program.[1] The school had expanded from a traditional small-aircraft aviation focus to drone education under Bergerson's initiative, not because of a system-level investment in urban aviation education, but because one educator recognized what was happening in the field and reoriented his classroom around it. The result is a program that is genuinely transformative for the students who find their way into it. It is also, by any honest measure, exceptional in the wrong sense: a program that should be standard is instead singular. Bergerson's students are fortunate. Their fortune is, in part, a function of their teacher's individual initiative, which is a fragile foundation on which to build a workforce pipeline.

The institutional infrastructure to do better already exists, and it is more powerful than is commonly recognized. There are 645 Hispanic-Serving Institutions (HSIs) — a federal designation under Title V of the Higher Education Act for institutions where at least 25 percent of enrolled full-time students are Hispanic in the United States in the 2024–25 academic year, including 226 two-year public institutions, 216 four-year public institutions, 11 two-year private institutions, and 192 four-year private institutions.[5] These are geographically distributed across precisely the regions where the drone industry's talent gaps are most acute: the Rio Grande Valley, the Central Valley of California, urban centers from Chicago to Miami, and rural communities across the Southwest. Many are located within commuting distance of the agricultural drone operations, infrastructure inspection corridors, and public safety agencies that are actively recruiting trained operators and technicians. The FAA's UAS Collegiate Training Initiative network, tracked by the National Center for

Autonomous Technology, includes a wide range of these geographically distributed institutions among its participating schools.[6] The architecture for a geographically distributed UAS talent pipeline is not missing. The investment to activate it at scale is.

The fundamental misalignment between where the jobs are emerging, where the talent lives, and where the industry has historically pointed its recruiting attention is not a coincidence. It is the residue of decades of recruitment patterns built around the geographies and institutions that already populated aviation and aerospace. Those patterns were never designed to exclude; they simply replicated themselves. Breaking the replication cycle requires deliberate effort: programs that go where the students are rather than waiting for the students to arrive, institutional partnerships that leverage existing community trust rather than trying to build it from scratch, and a clear-eyed understanding that the talent gap and the geographic gap in UAS workforce development are not two separate problems. They are the same problem, viewed from two different angles.

Planting the Seed Before the Talent Gap Becomes a Crisis

The workforce gap the UAS industry faces in 2026 was shaped by decisions, and non-decisions, made ten and fifteen and twenty years ago, in the classrooms and community centers and counselors' offices where young people first formed ideas about what careers were available to people like them. The workforce that will fill the industry's needs in 2035 and 2040 is being shaped right now, in those same settings. That is the essential logic of K–12 investment as a UAS workforce strategy: it is not primarily about the near-term talent shortage. It is about the long pipeline, about ensuring that the students currently in middle school and high school leave those years with a mental model of the drone industry that includes them.

The Dee Howard Foundation's work in San Antonio is worth examining in detail, because it represents a model with genuine structural ambition. More than twenty high schools in a single school district, Southwest ISD, running drone pilot training programs with direct pathways to FAA Part 107 certification. The programs are not electives in the marginal sense. They are designed to deliver the technical skills, the regulatory knowledge, and the flight simulation hours that prepare students for a federal commercial aviation credential. For students who are predominantly Hispanic and from low-income families, that credential represents a qualitatively different kind of economic opportunity than anything else currently available on the high school menu. The Foundation has done what the best philanthropic investments do: identified a specific community, a specific career pathway, a specific structural gap between the two, and built a bridge, at scale, within a single institutional system, with enough coverage to normalize drone careers as a realistic aspiration rather than an exceptional outlier.

The Center for Applied Special Technology's (CAST) Take Flight project operates at a different level of the system, middle school rather than high school, rural rather than urban, girls as the explicit target population, but with the same animating conviction: the message about who belongs in STEM must be actively changed, not merely passively hoped for. The NSF-funded program has reached more than 1,000 rural students across 75 or more schools in 11 states, training more than 100 teachers in a curriculum that takes drone technology as its multidisciplinary vehicle for expanding students' sense of what careers are available to them.[3]

The message about who belongs in STEM must be actively changed, not merely passively hoped for.

Take Flight deliberately reframes STEM goals in communal and altruistic terms rather than individualistic achievement terms. Research on STEM identity development, particularly for girls and for students from communities with limited aviation exposure, consistently shows that framing STEM as a path to personal achievement and individual distinction is less effective than framing it as a path to serving community needs. Drone technology maps naturally onto communal framing: precision agriculture serves farming communities. Search and rescue saves neighbors. Environmental monitoring protects the land. When a rural middle school student in North Dakota or Alabama hears that drone pilots are the people who help find missing hikers, map flooding damage for emergency responders, and monitor crop health for local farmers, the career becomes something other than a high-tech pursuit reserved for people from different zip codes. It becomes a way of being useful to the community she already loves.

The program also deploys strengths-based learning, building on what students already know, connecting drone skills to existing knowledge of local environments and local economies, rather than remediating presumed deficits. Teacher feedback from Take Flight participants captures the effect directly: "The students loved it and picked up the technical skills so quickly." Another educator reflected: "Without the curriculum, I would not be where I am today. I needed a jumping-off point."[3] That second observation matters as much as the first. The teacher pipeline is itself a constraint on K–12 drone program expansion. The FAA requires that high school drone pilot education programs be led by FAA-certified teachers. Building the teacher pipeline, supporting classroom teachers in earning their commercial remote pilot certificates, training them in evidence-based pedagogical approaches, connecting them to curriculum resources they don't have to develop from scratch, is therefore a prerequisite for program scale, not an afterthought to it.

The GAO's 2025 report on rural STEM education points consistently toward a single finding about what works for resource-constrained districts: the partnership model. Schools that successfully overcome the barriers of geographic isolation, thin staffing, and equipment scarcity do so by connecting with university research centers, nonprofits, and local employers, importing external resources into the school environment rather than expecting the district to build programs from scratch on a budget that cannot support them. Every successful K–12 drone program in communities the aviation pipeline has not historically reached has followed some version of the same logic: find the school, bring the partner, build the bridge, and trust that the students, when given the chance, will run across it.

The Institutions Already in the Field

The fastest path to closing the geographic pipeline gap is not to build new institutions from scratch. It is to work with the ones already embedded in the communities where the gap is widest. America's network of geographically distributed post-secondary institutions, regional community colleges, technical schools, and four-year universities in rural, agricultural, and post-industrial regions, already has what a workforce pipeline requires: enrollment pipelines, community relationships, and in many cases the geographic proximity to the precision agriculture, infrastructure inspection, and public safety drone operations that need trained workers. What many of these institutions lack is not the will or the student demand. It is the UAS curriculum, the flight equipment, the certified instructors, and the employer partnerships that turn credential programs into career pipelines.

The FAA's UAS Collegiate Training Initiative already reflects this logic. When institutions like Southwest ISD's partner colleges in San Antonio or the technical schools serving the rural communities documented in the GAO's 2025 report join the CTI network, the federal aviation education framework reaches students who would not naturally flow into a four-year aviation degree program at a major research university.[7] That connection is not incidental to the pipeline. It is the pipeline, a direct link between the federal program architecture and the geographic communities where drone adoption is accelerating fastest and trained workers are scarcest.

The funding to build these programs at scale is not absent. It is fragmented and, in some cases, newly uncertain. The Department of Defense directs STEM research and education funding to geographically distributed institutions as part of its broader effort to build a technically capable domestic workforce; drone-related engineering, systems integration, and technical training are natural extensions of that stream and it remains active. The NSF's Education and Human Resources directorate has historically funded programs that expand access to STEM careers, and a community college developing a UAS associate's degree with a demonstrated track record of placing graduates in commercial drone careers is exactly the kind of initiative that directorate was designed to support. However, the directorate has undergone significant restructuring since 2025, with more than 800 grants terminated and funding rates substantially reduced; organizations planning to pursue NSF education funding should verify current program availability directly before building strategies around it. The Economic Development Administration funds regional economic development investments in high-unemployment areas and a drone technology training program in a region where precision agriculture or infrastructure inspection is creating genuine job demand is a textbook EDA investment. EDA's FY2026 budget was reduced approximately 31 percent from prior year levels, narrowing but not eliminating the available funding pool. These funding streams exist. What is newly urgent is both the case for connecting them systematically to the UAS workforce pipeline and the organizational capacity to navigate a landscape that has become more competitive and less predictable than it was two years ago.

Any educator who has brought a drone into a classroom in a community that the aviation industry has historically overlooked will describe the same phenomenon: the interest is there. It has always been there, waiting in the communities the industry has not yet reached, unmet not because of any deficit of curiosity or ambition or aptitude but because the programs, the equipment, the certified teachers, and the industry partners willing to make hiring commitments had not yet arrived. Getting them there is the work of this moment.

The Trusted Intermediary

There is an assumption embedded in most workforce development strategy: that the school is the right entry point, that the pathway to a career begins in a classroom, runs through a formal credential, and ends with an employer. The assumption is not wrong, exactly, but it is incomplete in ways that matter enormously when the target population is young people whose relationship with formal educational institutions has been complicated by years of underinvestment, institutional neglect, and the accumulated weight of systems that were not designed with them in mind. In those communities, which are precisely the communities the UAS industry most needs to reach, the trusted intermediary is often not a school at all. It is a community center. A church. A youth program. A Boys & Girls Club chapter. A community development organization that has spent years building the relationships that schools and universities have not.

The most effective community-centered UAS workforce programs share a common architecture regardless of where they operate. They offer hands-on flight experiences in accessible settings, community centers, recreation facilities, school gyms, with no application process and no financial barrier to initial participation. They begin with exposure and move, for those who choose to continue, through education and credentialing to employment and entrepreneurship. Students come to these programs not because they are already inside a formal aviation pipeline, but because someone created a space for them in a setting they already trusted, and offered them an experience that was concrete and immediate and engaging in a way that a brochure about FAA certification is not. The experience often precedes the aspiration. The aspiration precedes the enrollment decision. The enrollment decision leads to the credential. The credential leads to the career. Disrupting that sequence, trying to recruit people into credentialing programs before they have had the experience that makes the credential meaningful, is one of the most common and most costly mistakes in workforce development strategy.

Research on career decision-making among students with limited prior exposure to an industry consistently shows that a concrete experience of the work, holding the controller, flying the aircraft, talking to a working professional, is a prerequisite for formal educational commitment, not a consequence of it. A student who has never held a drone controller and has never met a working drone professional from her community is not going to enroll in a UAS program based on labor market data alone. A student who has spent a Saturday morning flying at a community event organized by a trusted local organization, who has met a woman who looks like her mother and works as a drone data analyst for a utility company, and who has heard from someone she trusts that the Part 107 credential is achievable and that the minimum age to sit for the exam is sixteen, that student might. The pathway from zero to commercial drone pilot is genuinely one of the most accessible professional credentialing pathways in the American economy. That message exists. It is not yet delivered, with the consistency and the cultural attunement the moment requires, in the settings where the people who most need to hear it are most likely to receive it.

The economic framing of that message matters as much as its cultural delivery. In communities where four-year college completion rates fall below the national average, the argument that moves people is not "follow your passion in technology." It is "this is a pathway to a living wage that does not require four years of school and $60,000 of debt." Entry-level drone pilot salaries range from $45,000 to $65,000 annually. Experienced pilots earn between $65,000 and $120,000. Fleet supervisors and drone service company owners can earn from $100,000 to more than $180,000.[10] These are wages that represent genuine economic mobility for families in communities where the median household income has not meaningfully grown in a generation. The career is real. The income is real. The community-based trusted intermediary is, in many cases, the only institution positioned to deliver that message in a way that will actually be heard.

The Funding Is There. The Connections Are Not.

The policy architecture for building UAS talent pipelines in communities the industry has not yet reached is more complete than the industry's fragmented investment patterns would suggest. The Workforce Innovation and Opportunity Act (WIOA) was designed precisely for this moment: federal workforce development funding directed to youth and adults with significant barriers to employment, in high-demand occupations, delivered through community-level partnerships that can reach people who do not naturally flow into traditional workforce development systems.[9] UAS technician training at a community college serving students from a high-poverty urban neighborhood, delivered in partnership with a regional drone service company that has committed to hiring qualified graduates, is not a creative workaround to WIOA eligibility requirements. That is WIOA operating as designed.

The FAA invested $13.5 million in Aviation Workforce Development Grants in fiscal year 2024 across twenty-two individual grants, with explicit authorization for outreach and recruitment activities in communities not yet engaged by the aviation workforce pipeline.[8] NSF's Education and Human Resources directorate has historically funded programs like the CAST Take Flight project, specifically because the federal government has recognized that the STEM workforce national competitiveness requires is not going to materialize from the narrow slice of the population that has historically pursued aviation careers. Given recent restructuring and funding reductions at NSF, organizations pursuing this stream should confirm current program status directly. The Department of Defense directs STEM research and education funding to geographically distributed technical institutions through programs specifically designed to build workforce capacity in regions that traditional STEM pipelines do not reach, a stream directly applicable to UAS-related engineering and technical education.

The issue is not the absence of funding. It is the fragmentation of funding streams that were designed for adjacent purposes and have not yet been systematically connected to the specific opportunity that geographic UAS pipeline development represents. The K–12 drone program in San Antonio that produces Part 107-certified high school graduates is not automatically connected to the community college UAS degree program that could take those graduates to the next credential level. The community college that joins the FAA CTI network is not automatically connected to the drone service company forty miles away that is actively recruiting trained technicians and cannot find them. The community center that runs a Saturday drone program for middle schoolers is not automatically aware of the WIOA funding stream that could support expanding that program into a structured, multi-step career pathway with employer commitments at the end.

The communities with the highest youth unemployment and the least access to aviation careers are the same communities the industry most urgently needs. That alignment is not a coincidence. It is an opportunity.

Building the connections, between K–12 programs and community college credential pathways, between regionally embedded institutions and industry employers, between community-based exposure programs and federal workforce funding mechanisms, is coordination work. It is not glamorous work. It does not produce a single landmark program or a ribbon-cutting moment. It produces a pipeline: a connected sequence of experiences, credentials, relationships, and commitments that moves a young person from a Saturday morning drone flight at a community center to an employed professional career in commercial aviation. That pipeline does not build itself. It requires organizations that can see across the system, that know the students in San Antonio and the employers in Austin, the technical college in Huntsville and the drone manufacturer next door, the rural middle school in North Dakota and the precision agriculture drone operation one county over, and that have the relationships, the credibility, and the sustained presence to make the connections that produce careers rather than merely credentials.

Those organizations exist. Aerospace Career Pathways (ACP), LLC, a service-disabled veteran-owned business, and its nonprofit counterpart the Aerospace Career Pathways Foundation, build exactly this kind of connective infrastructure: workforce development programs that bridge youth, veterans, and communities into aerospace, aviation, and advanced technology careers.[11] The National Center for Autonomous Technology, which manages the FAA CTI program database and tracks institutional participation in the federal aviation education network, is another. Regional workforce development boards, community foundations, and state economic development agencies are others. What these organizations need is not a new program or a new policy framework designed from scratch. It is consistent, coordinated, adequately funded investment that connects the pieces already present into a pipeline that actually flows: from curiosity to exposure, from exposure to credential, from credential to employment, from employment to career, from career to the next Leonardo Fernandez who sees the number and decides that it has his name on it.

Looking Ahead — Chapter Eight

A pipeline is only as strong as what is waiting at its end. Programs build talent. Employers determine whether that talent finds careers. The best K–12 drone program in San Antonio produces graduates who need employers ready to hire them, develop them, and keep them. Chapter Eight is for those employers, and for an industry that cannot afford to lose the workers it has worked so hard to find.

Notes and Sources

1. Stewart Lawrence. "More High Schools and Colleges Are Adopting Drone Pilot Education Programs." LinkedIn, March 11, 2025. Source for Dee Howard Foundation / Southwest ISD programs; Leonardo Fernandez; Frankford High School Aviation Academy; Josh Bergerson enrollment data.

2. KSAT TV San Antonio, coverage of Southwest ISD drone pilot programs, cited in Lawrence (above), March 2025. Source for Rodolfo Urby quotation on opportunity and exposure for families in the district.

3. CAST. "Take Flight: Using Drones to Get Rural Middle School Girls Interested in STEM Careers." CAST.org, August 21, 2024. Source for program reach: 75+ schools, 11 states, 1,000+ students, 100+ teachers trained; program design principles; teacher feedback quotations.

4. U.S. Government Accountability Office. STEM Education: Selected Federal Initiatives, Challenges, and Approaches to Supporting Rural Populations. GAO-25-107371. July 15, 2025. Source for rural Alabama district with no certified math teachers for 300 students grades 7–12, 2023–24; transportation and logistics barriers; university/nonprofit partnership model findings.

5. Hispanic Association of Colleges and Universities (HACU). "2024–2025 HACU List of Hispanic-Serving Institutions." HACU.net, 2025. Source for 645 total HSIs by sector: 226 two-year public, 216 four-year public, 11 two-year private, 192 four-year private.

6. National Center for Autonomous Technology (NCAT). "UAS-CTI Schools & Programs." NCAT.org, accessed May 2026. Source for CTI network school listings, institutional type classifications, and geographic distribution of participating institutions across all regions of the country.

7. Federal Aviation Administration. "Approved UAS-CTI Schools Listing." FAA.gov, May 2025. Source for CTI program structure and FAA-certified teacher requirement for high school drone pilot education programs.

8. Federal Aviation Administration. "Funding Opportunities — Aviation Workforce Development Grants." FAA.gov, updated January 8, 2026. Source for $13.5 million in FY2024 Aviation Workforce Development Grants across 22 awards; explicit outreach and recruitment authorization.

9. U.S. Department of Labor. Workforce Innovation and Opportunity Act (WIOA) — Program Overview. DOL.gov. Source for WIOA eligibility framework: youth and significant-barrier-to-employment populations, high-demand occupations, community-level partnership delivery.

10. Military Transition Toolkit. Drone pilot salary data. January 19, 2026. Source for wage ranges: entry-level $45,000–$65,000; experienced pilots $65,000–$120,000; supervisors and owners $100,000–$180,000+.

11. Aerospace Career Pathways (ACP), LLC. acpathways.com. Service-disabled veteran-owned business; workforce development programs in aerospace, aviation, and advanced technology careers. Aerospace Career Pathways Foundation. acpathwaysfoundation.org. Nonprofit counterpart building equitable pathways for youth, veterans, and communities into aerospace and aviation careers.

The Employer's Playbook

"The future of uncrewed aviation will be built by employers who treat talent as infrastructure, not an afterthought."

— Sharon Rossmark

The Talent War Has Arrived

Call her Dana. Her official title is Operations Lead at a drone services company she helped build from the ground up eight full-time employees, twenty active contractors, infrastructure inspection work across the utility corridor. Three years ago, the founder bet the company's future on commercial UAS. The contract pipeline proved the bet right, growing faster than anyone projected. Dana helped build it. She also helped rebuild it after the 2023 round of layoffs that followed a funding shortfall and yet another BVLOS regulatory delay that pushed revenue projections out by eighteen months. That experience left its mark on how she thinks about hiring.

The two BVLOS operations specialist positions she has been trying to fill for the past three months are the most consequential hires she has ever attempted. The candidates exist. She has found them on LinkedIn. She can see their credentials, read their mission histories, identify the specific individuals whose combination of advanced airspace knowledge and systems integration experience maps almost perfectly onto what the company needs. But every one of them is already employed. Every one of them is fielding other offers. And at least one of them, a former drone delivery startup veteran who lived through two industry layoffs in eighteen months has made it plain that he is not joining any drone services company on a standard at-will employment basis without some form of contractual commitment to his role and runway. He has been burned before. He is not being unreasonable.

This is the talent war. It arrived quietly, without a press release or a formal announcement, at the intersection of a technology wave, a regulatory cycle, and a workforce that has learned through direct experience that the commercial drone industry's employment commitments are contingent in ways that traditional aviation and aerospace employers' commitments are not. Sixty percent of companies in the drone sector now cite a lack of qualified applicants as their primary barrier to growth, not capital constraints, not customer demand, not technology readiness.[1]

The employer who figures out how to win the talent competition is not just solving an HR problem. They are building the competitive moat that will define their position for the next decade.

This chapter is for the employer. It is a practical, evidence-based guide to hiring, retaining, and developing UAS talent in the most competitive market the industry has ever seen. It begins with an honest assessment of the structural conditions the industry has created for itself because no talent strategy succeeds in a context it refuses to acknowledge.

The Contractor Architecture: How the Industry Actually Works

The Reality Behind the Org Chart

Most drone services companies do not look like the workforce charts in aviation industry reports. They look like Dana's company: a small core of full-time employees surrounded by a larger constellation of contractors, licensed Part 107 pilots, UAS technicians, data processing specialists engaged per-project, paid per-deliverable, and released between contracts. This is not a failure of ambition. In a market where revenue is contract-dependent, regulatory timelines are uncertain, and the capital cycle has repeatedly proven unable to sustain expanded payrolls through BVLOS delays, contractor-based operations are rational capital management.

The Drone Industry Insights Global State of Drones 2025 report found that 48 percent of commercial drone companies worldwide still have fewer than ten employees, while the share with 50 to 200 employees grew to 37 percent a meaningful shift, but one that confirms the overwhelming majority of drone services firms are small businesses navigating the contractor-to-employee transition in real time, not enterprise organizations with dedicated HR infrastructure.[13]

Understanding the contractor model as a spectrum not a binary choice is the first requirement of an honest workforce strategy for this industry. At one end is the strategic contractor relationship: a skilled professional engaged repeatedly across multiple contracts, treated as a reserve workforce member, compensated fairly, kept informed about the company's pipeline and direction, and on a defined path toward full-time conversion when the contract volume justifies it. At the other end is the aggregator-mediated gig deployment, which represents a significant and largely invisible layer of how commercial drone work actually gets done in the United States and which deserves the industry's honest attention.

The Gig Floor: Aggregators, Piece Rates, and the Pilot Who Pays the Bill

Behind the market projections and the BVLOS investment narratives is a sector of the drone workforce that rarely appears in industry reports. They are the contract pilots who work through job aggregator platforms that connect clients needing drone services with licensed Part 107 operators available to perform them. A solar energy company needs panel inspections across a 200 acre installation. An insurance carrier needs roof damage documentation across 150 homes in a hurricane impact zone. A billboard network needs proof-of-posting documentation across a regional corridor. The aggregator takes the client relationship, sets the scope, defines the deliverable criteria, and deploys a pilot network to execute.

The pilot is at the bottom of this food chain in ways that deserve to be named plainly. In the disaster deployment scenario that has become a recurring feature of drone industry work post-hurricane, post-tornado, post-wildfire the dynamics are particularly stark. A pilot contracted to document roof damage across a designated grid of homes may be assigned 100 or more properties at a specified rate per capture, subject to defined quality criteria. From that per-capture payment, the pilot absorbs every operational cost: fuel, lodging, equipment wear, liability insurance, the time spent planning flight paths and processing deliverables, and the logistical reality that in a disaster zone, nothing is convenient. The hotels nearest the impact area are fully occupied by displaced residents who had nowhere else to go. Gas stations close to the damage corridor have run dry or are charging elevated prices. The pilot drives further, pays more, and earns less per effective hour than the per-capture rate suggests when the full cost picture is calculated.

The pilot who documents a hundred damaged roofs in the wake of a hurricane is doing work with genuine social value. The economics of how that work is structured rarely reflect it.

This is not an indictment of aggregators as a business model. Aggregator platforms perform a genuine market function: they create accessible, repeatable work for licensed pilots who would otherwise spend their own time and resources finding clients, and they provide clients with scalable, compliant drone services without the overhead of building internal programs. But the economics of the gig floor have workforce development consequences that compound over time. Pilots working primarily on piece-rate aggregator deployments are not building institutional relationships, the operational depth, or the advanced credentials that the industry needs at the BVLOS scale. They are providing a service. They are not building a career. And the attrition rate from gig-tier drone work into adjacent fields traditional aviation, construction inspection, geospatial services is one of the industry's least-discussed talent pipeline leakages.

The Conversion Decision: When a Contractor Becomes an Employee

The most consequential and most anxiety-producing workforce decision most drone services company leaders face is not which candidate to hire. It is when to convert a contractor they already trust into a full-time employee they are financially committed to. The question is not philosophical. It is a cash flow calculation with significant fixed-cost implications, and getting it wrong in either direction is expensive.

Converting too early before the contract backlog is deep enough and stable enough to support a fixed payroll commitment puts the organization in the position of carrying employee overhead through the gaps between contracts that every drone services firm experiences. Converting too late keeping a high-performing contractor in contractor status after they have earned the organizational trust and role continuity of a full-time employee creates both a legal exposure and a retention risk. The IRS and the Department of Labor apply specific tests to determine whether a worker classified as an independent contractor is actually functioning as an employee; misclassification carries penalties and back-tax liability that have cost drone companies money they did not expect to spend.

The triggers that reliably signal readiness for conversion include: a contract backlog of sufficient depth that the employee's fully-loaded cost is covered across at least two contract cycles without assuming new business; a role that requires continuity of institutional knowledge, client relationship management, or operational authority that cannot be effectively transferred to a different contractor between engagements; and a contractor whose departure to a competitor or to a full-time role elsewhere would materially damage the organization's operational capability or client relationships. When all three conditions are present simultaneously, the conversion decision has already been made by circumstances. The only question is whether the organization will make it on its own terms or be forced into it by the contractor's departure.

The fully-loaded cost of a first full-time UAS hire is consistently underestimated by first-time employers in this sector. Base salary is the most visible component. Add employer-side payroll taxes (approximately 7.65 percent of base), workers' compensation insurance (elevated for field operations roles), health benefits (averaging $6,000 to $8,000 annually per employee in 2026), equipment and tool provisioning, onboarding time, and the administrative overhead of HR compliance, and the true annual cost of a $65,000-base field operations role is closer to $85,000 to $90,000. Employers who budget to base salary and then encounter the additional costs midway through the first year are making an estimation error that better planning would have prevented.

The industry cannot build the talent depth it needs on a foundation of piece-rate gig work and perpetually deferred employment commitments. That is not a moral argument. It is a workforce supply chain argument.

The Trust Deficit and What It Costs

The commercial drone industry has a trust problem with some of its most capable people, and it created that problem itself. The capital raise → hire up → regulatory delay → cash constraint → layoff → contractor surge cycle has repeated often enough across enough companies — from well-funded drone delivery startups to regional inspection firms — that experienced UAS professionals have internalized the pattern. They have seen colleagues laid off after relocating for drone industry positions. They have watched funded companies run out of runway waiting for BVLOS clarity. They have received recruiting calls from companies that were in a hiring freeze six months later.

This trust deficit shows up in recruiting conversations as the specific, concrete reluctance that Dana encountered: the experienced BVLOS specialist who will not join on standard at-will terms, the autonomy engineer who adds an equity requirement to a compensation negotiation because salary alone does not offset the perceived employment risk, the data analyst who chooses the aerospace prime over the drone services firm offering more interesting work because the prime's employment stability is not in question. These are rational responses to documented industry behavior. Employers who want to win these candidates must address the trust deficit directly — not with platitudes about company culture, but with structural commitments: defined contract terms, transparent communication about runway and regulatory dependencies, clear conversion pathways from contractor to employee, and in some cases equity participation that converts employment risk into upside participation.

The Six Roles That Will Define Your Competitive Position

Know What You're Hiring For

The first and most consequential step in any serious UAS talent strategy is accurate role definition, and most employers are getting this wrong in ways that cost them months of recruiting time and thousands of dollars in opportunity cost before anyone fully diagnoses the problem. The UAS job market has changed fundamentally from its early Part 107 years. In 2019, the dominant hiring need was Part 107-certified drone pilots for visual line of sight inspection and photography work. Today, the talent landscape has fractured into at least six distinct role categories where demand consistently outpaces supply, each with its own credential requirements, compensation expectations, career trajectory, and accessible talent pool. An employer who approaches all six as variations of the same "drone professional" job will fail to recruit any of them effectively.[3]

BVLOS Operations Specialists represent the rarest and most urgently needed credential in the commercial UAS market. These are professionals who can design, plan, and execute beyond visual line of sight (BVLOS) operations, a function that sits at the intersection of advanced airspace knowledge, systems integration capability, and mission execution discipline. With the FAA's Part 108 BVLOS framework in its final regulatory stages, every enterprise drone program and every drone services company anticipating growth will need this capability as operations scale. Almost none of them can find it. Recruiting a BVLOS specialist requires not a standard job posting but a fully developed value proposition: the employer who wants to win this candidate must be able to articulate clearly what their BVLOS program will allow this professional to do that they cannot do anywhere else. The salary band is the table stake. The value proposition is the differentiator.

Autonomy and Artificial Intelligence (AI) Engineers are developing the machine learning models that allow drones to navigate without continuous human input: flight path optimization, object detection, obstacle avoidance, and fully autonomous mission decision-making. This is the role category most aggressively competed for across the entire technology sector. A drone company posting for an autonomy engineer is competing with Waymo, Apple, and every defense prime in the country for candidates from the same constrained talent pool. Compensation at the top of the range for autonomy engineers reaches $150,000 or more, with equity packages that can multiply that base figure significantly. The autonomy engineer who wants to work on something that genuinely matters — in a domain where individual contributors can still shape system architectures rather than optimize sub-components of sub-systems — is a candidate the drone industry can attract and win. But only if the employer has thought clearly about why their specific work environment offers something the large technology company cannot.

UAS Systems Integrators bridge the gap between hardware, software, sensor suites, and data platforms, building the operational systems that enterprise clients deploy in the field. This role is frequently filled from adjacent engineering disciplines: aerospace engineers, electrical engineers, mechatronics specialists who bring a foundation of technical knowledge that translates directly into UAS integration work. The talent pool is broader than for pure autonomy engineers, which is an advantage. But the credential confusion that plagues this category — no standardized UAS systems integrator certification currently exists — makes it genuinely difficult for employers to evaluate candidates consistently. Employers who define this role with precision, specifying the exact hardware environments, software platforms, and sensor systems the integrator will work with, will recruit more efficiently and evaluate candidates more accurately than those who rely on general engineering credentials as proxies for UAS-specific readiness.

Data and Geographic Information Systems (GIS) Analysts process the imagery, LiDAR (Light Detection and Ranging) point clouds, and sensor data that drones capture and turn raw capture into actionable intelligence for clients. This is one of the most accessible high-value UAS roles for adjacent-field professionals, and employers who recognize this have a materially wider talent pool than those who hire only from traditional UAS backgrounds. The GIS analyst, the environmental scientist, the construction project manager who already understands site documentation and spatial data: each needs a specific, bounded set of additional skills to become productive in a drone data role, not a complete professional reinvention. Infrastructure inspectors who reskill as drone data reviewers save their employers 400 hours of field work annually, a figure that transforms the business case for upskilling investment from a training expense into a productivity calculation.[4]

Drone Compliance and Regulatory Affairs Specialists have moved from the periphery of UAS organizational charts to the center of operational risk management in the span of three years. NDAA procurement compliance, Blue UAS certification, FAA airspace authorizations, Remote ID compliance, export control requirements, and the evolving BVLOS regulatory framework all require dedicated professional attention that an operations manager with seventeen other responsibilities cannot reliably provide. This is the role most frequently underhired in the drone sector, treated as a collateral duty until the compliance failure, the lost federal contract, or the FAA enforcement action makes the case for dedicated headcount in the most expensive way possible.

Counter-UAS Specialists, professionals who detect, identify, and neutralize unauthorized drone activity, represent one of the fastest-growing defense-adjacent roles in the market. The proliferation of drones across civilian, commercial, and adversarial domains has created genuine, urgent demand from the Department of Defense (DoD), the Department of Homeland Security (DHS), law enforcement agencies, airport authorities, and critical infrastructure operators. For employers serving defense and critical infrastructure clients, counter-UAS capability has moved in the past 24 months from a differentiating service offering to a baseline client expectation.

What the Market Is Actually Paying

The Compensation Reality

The employer who has not run a compensation analysis against current market data in the past twelve months is almost certainly underpaying. And the people who know they are underpaid are always the first to leave when the recruiter calls. The UAS market has been moving rapidly enough in the past two years to have left a significant number of employers operating with structures that are no longer competitive, even if they were accurate eighteen months ago.[3]

Glassdoor data as of mid-2026 places the average commercial drone pilot salary at approximately $94,700, with top earners at the 90th percentile reaching $173,000 annually, a figure that surprises many employers who entered the drone market expecting to hire operators at rates closer to the early Part 107 era, when a brief oversupply of newly certified pilots temporarily depressed wages. That era is over. Leading employers are currently offering 10 percent annual salary increases to retain certified operators in a market where lateral recruitment is constant, and the candidate has real options. The 55 percent of HR managers in the drone industry who cite pilot shortages as their top recruitment barrier are not describing a problem that more job postings will solve. They are describing a compensation reality that more accurate benchmarking will begin to address.[12]

Table 8.1 — UAS Role Compensation Benchmarks, 2026

Role	Salary Range	Notes
Commercial drone pilots/operators	$50,000–$100,000+	Avg. $95,000; leading employers offering 10% annual increases
Hardware and systems engineers	$83,000–$155,000	Varies by specialization and systems environment
Software and autonomy developers	$85,000–$150,000	Median $120,000; 12% bonus typical
Data and GIS analysts	$55,000–$90,000	20% salary increase with multispectral imaging mastery
Field technicians	Avg. $85,000 + $10,000 hazard pay	Growing 25% annually through 2028
Sales and business development	$60,000–$120,000+	Frequently undercompensated in UAS organizations
Training and curriculum development	$50,000–$80,000	Growing with program expansion across sector

Sources: Glassdoor (commercial drone pilot compensation, mid-2026); DroneBundle, Gitnux, and WifiTalents, 2026 (industry-reported estimates for other UAS roles). Figures reflect industry-reported survey data rather than U.S. Bureau of Labor Statistics data, as the BLS does not yet track UAS-specific occupational titles separately from related fields.

The benefits landscape has also standardized at a level that many smaller drone employers have not yet reached. Eighty-five percent of U.S. drone firms now offer 401k matching up to 6 percent. Ninety-two percent provide flight insurance, a non-negotiable for employees who operate aircraft professionally. Sixty percent offer stock options as a retention mechanism. Bonuses for certified drone specialists average 15 percent of base pay.[2]

One of the most significant and systematically underappreciated compensation elements in UAS talent strategy is the value of training investment as a component of total compensation. The employer who funds BVLOS training, at an average cost of $5,000 per employee, is making a bet that pays in multiple currencies simultaneously: in the expanded operational capability of the trained professional, in the unambiguous retention signal that this organization invests in professional growth rather than simply extracting it, and in the market premium that a BVLOS-capable operator commands when the regulatory window opens. LiDAR specialization adds an average of $50 per hour to a drone pilot's market rate. Multispectral imaging mastery produces a 20 percent salary increase in agricultural drone roles.[4]

Why Your Best People Are Being Recruited Right Now

Retention: The War You Win Before It Starts

In a market where the 45-day average time to hire has become the optimistic scenario for specialized UAS roles, the arithmetic of turnover is unforgiving. Every position that opens costs an employer a minimum of six weeks of lost productivity before a replacement begins work, followed by a ramp-up period that can extend three to six months for roles requiring deep operational knowledge and client relationship continuity. The true cost of turnover in a skilled UAS professional role, counting recruiting fees, lost productivity, onboarding investment, and the institutional knowledge that walks out the door, is typically 1.5 to 2.5 times the annual salary of the departing employee.[2]

> ***The people most likely to be recruited away are the most capable and most credentialed — exactly the people the employer can least afford to lose.***

The retention challenge in UAS is compounded by the same market dynamics that make hiring difficult in the first place. The professionals most attractive to external recruiters are precisely the ones an organization can least afford to lose: BVLOS specialists who carry operational authority no one else on the team can exercise, autonomy engineers who are three months into an architecture that exists nowhere outside their own understanding, systems integrators with deep enterprise client relationships that are partly personal rather than purely institutional. These professionals receive constant inbound recruitment interest not occasionally but continuously and their loyalty is earned through career development that is real rather than rhetorical, through work meaningful enough to justify not pursuing other offers, and through organizational clarity about where the company is going and what specific role they will play in getting there.

Career ladder design is the most underutilized retention tool in the UAS employer's arsenal. Seventy percent of drone pilots are currently self-taught or certified through online programs; many have no clear picture of where their career goes from the position they currently hold.[4]

The employer who can show a drone operator a mapped career trajectory from Part 107 operator to BVLOS specialist to fleet operations manager to director of UAS operations, with clear milestones, credential requirements, and compensation progression at each level is giving that professional something the recruiter calling on Tuesday genuinely cannot offer: a future inside the organization. Career clarity costs almost nothing to create. It is one of the most durable retention mechanisms available.

Mentorship programs deserve particular attention because the retention case for them is directionally clear. New hires who have a structured mentorship relationship are more likely to stay, more likely to develop institutional knowledge that makes them genuinely productive, and more likely to recruit peers from their own professional networks. Women grew their share of Part 107 remote pilot certifications by 19.3 percent in 2025.[11]

Veterans bring operational discipline and mission-execution experience that the commercial drone market urgently needs and cannot replicate through conventional training. Generation Z represents a growing share of entry-level drone roles, bringing digital fluency and technology comfort that employers need but retaining them requires formal professional development structures that make career investment a sustainable reality rather than a recruitment talking point.

The Pipeline You Build Yourself

Building from Within: The Training Imperative

Employer-aligned training produces its strongest results when it is designed around operational outcomes rather than credential generation. The most effective enterprise UAS training programs begin with a foundational assumption that many providers do not make: that employees do not arrive mission-ready, and that the training investment must bridge that gap rather than assume it away. Programs structured around this principle combine online and in-person delivery, embed custom standard operating procedure development directly into the curriculum, and produce graduates who hold not merely certificates but procedures and institutional knowledge. That last element matters because it multiplies the return on the original training investment: a professional trained to the organization's own SOPs can train colleagues, compounding the value of the initial commitment in ways that single-credential programs cannot replicate.

The 12-week UAS maintenance technician program model developed by the Unmanned Safety Institute (USI), in partnership with Central Arizona College, Horry-Georgetown Technical College, and West Georgia Technical College, is not a theoretical construct. It is a deployed, operational training pathway built in partnership with employers who helped design the curriculum and committed to hiring the graduates.[8]

That last element is not incidental. Employer commitment to hiring graduates transforms a training program from a charitable workforce development initiative into a supply chain management strategy applied to human capital, and it is what makes community college and technical school partners willing to invest in developing the curriculum in the first place.

The absence of a regulatory baseline for drone technician qualifications there is currently no UAS equivalent to the FAA's Airframe and Powerplant (A&P) certificate means that employers are effectively on their own in defining what maintenance competency looks like for their specific fleet, their specific operational environment, and their specific mission profiles. ASTM International's F3600 UAS Maintenance Technician Standard provides a framework, but adoption is not uniform, and employers who have not engaged with it are making hiring and training decisions in the absence of the quality benchmarks that professional standards exist to provide.[9]

Reskilling from adjacent fields is the highest-volume, fastest-time-to-productivity strategy available to most UAS employers, and it remains systematically underused relative to its potential. The surveyor who already understands geospatial data needs photogrammetry software training and drone operations certification, not a full career reorientation. The construction project manager who already understands site documentation needs drone data platform training, not a four-year aviation degree. The traditional aviation mechanic who understands aircraft systems and FAA maintenance documentation needs UAS-specific systems knowledge to transfer their professional foundation into a domain where that foundation has immediate and high value.

The upskilling priority employers must address with the greatest urgency is sensor integration and data analysis, identified by 45 percent of current drone operators as their top development need.[4] The operator who can fly competently but cannot process, interpret, and deliver the data that the drone's sensors capture is providing roughly half the value that the technology is capable of producing for the client. Closing that gap through targeted training, tool access, and mentorship with data professionals does not require hiring a new employee. It requires investing in the professional development of someone the organization already knows, already trusts, and has already paid to develop into an operational asset.

Your Talent Strategy Self-Assessment

The Six Questions Every UAS Employer Must Answer

At the end of every strategy conversation, the question that matters is not what the employer believes about their talent strategy. It is what they can demonstrate. The six questions that follow are diagnostic: the employer who cannot answer them clearly and honestly does not have a talent strategy. They have a talent hope, which is a different and far more expensive thing.

The first question: does the organization know precisely what skills and credentials each UAS role on the team requires not generically, but specifically for the missions clients need today and the BVLOS operations they will need in 18 months? The employer who cannot answer this is recruiting against a vague profile and hiring people who will need to be substantially retrained before they are productive in the role. Role definition is not an HR function. It is a strategic function.

The second question: is the organization's compensation structure benchmarked against current market data? Not the survey from 18 months ago, not the hiring manager's intuition about what "the market" looks like, but the actual, current market in which leading employers are offering 10 percent annual salary increases to retain certified operators who have real options. The employer who believes their compensation is competitive based on outdated benchmarks is not being fiscally prudent. They are operating with false confidence in a market that has moved on without them.

The third question: can every UAS professional on the team describe their career trajectory within the organization in specific, concrete terms? Not "there will be opportunities as we grow," but an actual mapped career ladder with clear milestones, credential requirements, and compensation progression at each level. The drone operations professional who cannot see a defined path from where they are to where they want to go will eventually find an organization that provides that clarity because the recruiter calling on Tuesday will make a point of offering it.

The fourth question concerns BVLOS readiness: not if the organization will need this capability, but when, and whether the investment in building it is already underway. The employers who have already invested in BVLOS training at $5,000 per key employee will be operationally ready within days of the rule's finalization. The employers who have not will spend the following twelve months scrambling to credential the team they should have prepared eighteen months earlier, while their BVLOS-ready competitors capture the contracts that will define the enterprise drone services market for the next decade.[7]

The fifth question: how many UAS professionals on the team participated in a formal mentorship or professional development program in the last twelve months? For most employers, the honest answer is fewer than half. This is the employer who is investing in recruiting and underinvesting in the retention of what recruiting produces. Mentorship programs that meaningfully improve new hire retention are a foundational talent strategy tool for any employer who cannot afford to lose the professionals it has worked hard to find, and the cost of running a genuine program is a fraction of the cost of the turnover it prevents.

The sixth question and the one most specific to this industry's structural reality: does the organization have a documented contractor-to-employee conversion strategy? Does it know, specifically, which contractor relationships represent the highest-priority conversion candidates — and at what contract volume threshold, and under what runway conditions, that conversion becomes both financially defensible and strategically urgent? The company that cannot answer this question is making its most consequential workforce decisions reactively, in response to a contractor's departure notice or a client's demand for operational continuity guarantees. Those are the most expensive conditions under which to make a hiring decision. The time to answer this question is before the conversation becomes urgent.

The employer who can answer all six questions clearly and honestly has a talent strategy. The employer who cannot is operating on hope — and in the most consequential hiring period the drone industry has ever seen, hope is not a strategy.

Looking Ahead — Chapter Nine

Even the most expertly executed employer talent strategy operates within a larger ecosystem that no single organization controls. The scale of the commercial UAS market and with it, the scale of the talent opportunity is fundamentally determined by the public and private capital flowing into uncrewed aviation: the federal research investments, the state economic development commitments, the defense procurement decisions, and the infrastructure appropriations that together determine how large the drone economy will become and how quickly. Understanding where that money is going, where it is not going, and what it means for the employers and workforce developers building for this industry is the subject of Part III. Chapter Nine follows the money.

Notes and Sources

1. ABI Research. Commercial drone market projection to $58 billion by 2026, as cited in Christian & Timbers, March 2026. U.S. cited as largest regional market share. The 60 percent barrier-to-growth figure reflects industry-reported survey data (Gitnux, HR in the Drone Industry Statistics 2026), not U.S. Bureau of Labor Statistics or other government data.

2. Gitnux. "HR in the Drone Industry Statistics 2026." Gitnux.org, updated 2026. Source for compensation benchmarks, benefits data, 45-day average time to hire, and workforce composition figures throughout this chapter. These figures reflect industry-reported survey data rather than U.S. Bureau of Labor Statistics data, as the BLS does not yet track UAS-specific occupational titles separately from related fields.

3. Jeff Christian. "2026 Drone Industry Hiring Trends & Talent Strategy in the US." Christian & Timbers Insights, March 31, 2026. Primary source for role category taxonomy and market overview.

4. WifiTalents. "Upskilling and Reskilling in the Drone Industry Statistics 2026." WifiTalents.com, February 12, 2026. Source for skills gap data, self-taught pilot statistics, maintenance role growth projections, and reskilling ROI figures. These figures reflect industry-reported survey data rather than U.S. Bureau of Labor Statistics data, as the BLS does not yet track UAS-specific occupational titles separately from related fields.

5. DroneBundle / Jessica May. "Drone Jobs: Complete Guide to Careers in the UAV Industry 2026." DroneBundle Blog, January 24, 2026. Source for salary ranges across role categories and compensation structure data.

6. Aviation International News / James McDanolds, Sonoran Desert Institute School of Uncrewed Technology. "Drone Industry Faces Talent Crunch Ahead of BVLOS Launch." April 21, 2026. Source for talent pipeline analysis and regulatory readiness context.

7. Federal Aviation Administration. Notice of Proposed Rulemaking: Operation of Unmanned Aircraft Systems Beyond Visual Line of Sight (Part 108). Published August 7, 2025. Executive Order 14307 directed the FAA to finalize the rule within 240 days of June 6, 2025, establishing a target of approximately February 1, 2026. A 43-day government shutdown beginning October 1, 2025 — the longest in U.S. history — extended the target date to approximately March 16, 2026. The FAA additionally reopened a limited comment period on electronic conspicuity provisions through February 11, 2026. As of June 2026, the final rule has not yet been published; finalization and implementation remain pending.

8. Unmanned Safety Institute (USI). UAS Maintenance Technician Program launch announcement. February 2026. Source for 12-week technician training program model and community college partnership details including Central Arizona College, Horry-Georgetown Technical College, and West Georgia Technical College.

9. ASTM International. F3600 Standard for UAS Maintenance Technician. Reference for maintenance competency framework discussion; adoption status as of 2026.

10. Federal Aviation Administration. Civil Airmen Statistics, December 31, 2025. FAA Aeronautical Center, Oklahoma City, OK. Source for remote pilot certificate issuance data and gender-based certification trends.

11. Federal Aviation Administration. Civil Airmen Statistics, December 31, 2025. Source for the 19.3 percent year-over-year growth in women's Part 107 certification rate cited in the retention section.

12. Glassdoor. "Drone Pilot Salary in United States." Glassdoor.com, salary data as of May 2026. Average salary $94,699; 25th percentile $71,024; 75th percentile

$131,310; 90th percentile $173,299. Based on 230 salaries submitted anonymously by Drone Pilot employees in the United States.

13. Drone Industry Insights. Global State of Drones 2025 Report. Hamburg: Drone Industry Insights, 2025. Survey of 768 respondents across 87 countries, conducted mid-2025. Company size distribution: 48.2% fewer than 10 employees; 37% between 50 and 200 employees.

PART III: FOLLOWING THE MONEY

Future Flight

How Talent, Capital and Manufacturing will Shape the Next Era of Uncrewed Aviation

About Part III

Having mapped the landscape of uncrewed aviation in Part I and prescribed the workforce strategies in Part II, this section turns to the capital layer, the investment flows, public and private, that will ultimately determine whether the industry's ambitions are achievable. Part III follows the money across three distinct channels: public investment through federal agencies and state governments, private capital flowing into the drone startup ecosystem, and defense spending that functions simultaneously as national security priority and industrial policy. A fourth chapter examines how regional economies can position themselves to capture the long-term economic value the industry creates. Together, these chapters answer the question that every entrepreneur, executive, and policymaker in the uncrewed aviation space is asking: where is the money going, and is it going to the right places?

Follow the Money

"Public investment doesn't follow innovation — it accelerates it. Uncrewed aviation will be no exception."

— Sharon Rossmark

The Airport Director's Inbox

On a Monday morning in early spring, the director of a mid-sized regional airport in the American heartland sits down at her desk and opens her email. The airport she manages, a public-use facility with a single paved runway, a modest terminal, and an apron that has become an informal staging ground for drone operations serving the surrounding agricultural counties, is not a major hub in any conventional sense. But in the emerging geography of uncrewed aviation, it has become something increasingly important: a local node in the national drone ecosystem, a place where the federal government's expansive ambitions for UAS integration meet the operational realities of the communities that will actually live with those ambitions.

Her inbox, on this particular morning, contains four separate items from four separate federal agencies. The Federal Aviation Administration (FAA) has issued an invitation for applications under its Aviation Workforce Development Grant program, the same program that distributed $13.5 million across 22 grants in fiscal year 2024, explicitly authorized for the recruitment of UAS operators and aviation maintenance workers through education, outreach, and community partnerships.[11] The National Aeronautics and Space Administration (NASA) has sent a notice about a research partnership opportunity related to UAS Traffic Management (UTM) key site participation. The Economic Development Administration has circulated a solicitation for regional innovation grants with language that maps closely onto the drone corridor development strategy her airport's board approved two years ago. And the Department of Labor has published a Workforce Innovation and Opportunity Act (WIOA) funding notice that speaks directly to the drone technician apprenticeship program she has been trying to launch for eighteen months.

Each program is real. Each represents genuine federal investment in exactly the infrastructure she is trying to build. Each has its own application portal, its own eligibility criteria, its own definitions of what constitutes a successful outcome, its own program officers who operate in organizational silos so complete that none of them appear to be aware the others exist. She does not have a full-time grants manager. She has an operations assistant, a board that meets quarterly, and a strategic plan that identified UAS as a regional economic development priority two years before most of this money came into existence.

She is not ungrateful. She is overwhelmed, and she is representative. The regional airport director is not an edge case or an outlier. She is the central actor in the local UAS ecosystem that federal investment is designed to build: the administrator who manages the infrastructure, supervises the airspace coordination, attracts the operators, and connects the workforce pipeline to the industry. What the federal government has done, with the very best intentions, is design an investment strategy that imposes the highest coordination costs on exactly the actors who have the least capacity to bear them.

This is the animating tension of this chapter. The United States has made extraordinary public investments in uncrewed aviation, through legislation, executive action, agency programs, and state-level initiatives that collectively represent billions of dollars committed to the industry's growth. The question this chapter asks is not whether the investment is real. The money is real. The intent is genuine. The question is whether the architecture through which the investment flows is capable of producing the national industrial leadership that the legislation, executive orders, and press releases have promised.

The Summer of Appropriations

To understand the scale of what the federal government committed to uncrewed aviation in the summer of 2025, it helps to understand what had come before. For most of the decade preceding it, federal UAS investment had been characterized by a cautious, incremental quality, pilot programs, research grants, test site designations, regulatory studies, that reflected an agency culture understandably focused on managing the risks of a new technology rather than accelerating its deployment. The FAA, constitutionally an agency of deliberation and process, moved at an FAA pace. NASA's UTM research program was rigorous and foundational but operated on academic timelines. Congressional appropriations for drone-specific programs existed but were modest in scale. The federal government was watching the drone industry develop, occasionally nudging it, but not yet leading it. That changed in the summer of 2025 in a way that the industry had not seen before and may not see again for some time.

On June 6, 2025, the White House issued an executive order (EO) titled "Unleashing American Drone Dominance." The title alone marked a departure. Previous administrations had used cautious, regulatory language when discussing UAS policy. The word "dominance" carries a different charge: it is the vocabulary of industrial competition, of national strategic ambition, of a government that has decided it is no longer content to manage a technology and intends instead to lead it globally.[1] The EO directed all federal agencies to prioritize American-made drones in procurement decisions. It called for the streamlining of certification and approval processes across the federal government. It directed the promotion of American UAS exports to international markets, an explicit acknowledgment that the drone industry is a global competition and that American manufacturers need the federal government as a commercial ally, not merely a regulator. And it set a thirty-day deadline for the FAA to issue a proposed rule for BVLOS operations, an extraordinary pressure on an agency whose rulemaking processes typically measure in years, not months.

The FAA responded. On August 7, 2025, the agency published its Part 108 BVLOS Notice of Proposed Rulemaking, a rule that, when finalized, will unlock the commercial market for long-range drone operations that entrepreneurs, investors, and operators have been anticipating for years. Approximately 3,100 public comments were submitted, reflecting the depth of stakeholder engagement from the operators, manufacturers, and service providers who have been working toward this regulatory milestone. Implementation is expected in late 2026 or early 2027, but the direction is now clear: BVLOS at commercial scale is no longer a question of whether but of when.

The One Big Beautiful Bill Act (OBBBA), signed on Independence Day 2025, translated executive ambition into legislative appropriation at a scale the industry had never seen. The OBBBA's drone and defense provisions were not aspirational: they were specific enough to read as a production investment plan.[2] The act directed $1.4 billion to scale the small unmanned aircraft systems (sUAS) industrial base, a commitment designed to close the gap between the demand signal created by years of National Defense Authorization Act (NDAA) procurement restrictions on foreign-made drones and the domestic manufacturing capacity needed to answer that demand. It provided $250 million for a digital test environment for autonomy validation, not a research program but a production-scale testing infrastructure. It included $2.1 billion for medium uncrewed surface vessel development and $188 million for maritime robotics research and development, investments that signal the federal government's intention to extend drone dominance beyond the airspace and into the maritime domain.

The word "dominance" carries a different charge: it is the vocabulary of industrial competition, of national strategic ambition, of a government that has decided it is no longer content to manage a technology and intends instead to lead it globally.

The FAA Reauthorization Act of 2024, Public Law 118-63, enacted the previous year, provided the regulatory and programmatic scaffolding upon which the 2025 executive actions would build. Section 925 established a $6 million annual appropriation, running from fiscal year 2025 through fiscal year 2028, for the FAA's seven designated UAS test sites, funds structured as matching funds for commercial entities contracting with a test range to demonstrate or validate technologies essential to safe UAS integration.[3] The Reauthorization Act also provided the authorization basis for the FAA's Aviation Workforce Development Grant program. These were not large numbers by federal budget standards, but they were real, recurring appropriations, the kind of sustained annual commitment that allows institutions to build programs rather than chase one-time grants.

Taken together, the legislative and executive actions of 2025 represent the most coherent and financially serious federal commitment to uncrewed aviation the industry has seen. The question of whether the 2025 investment surge produces the industrial transformation it promises will be answered not in the press releases of 2025 but in the production numbers of 2028 and the market share statistics of 2030.

Table 9.1 — Key Federal Legislative and Executive UAS Investments, 2024–2025

Legislative / Executive Action	Year	Key UAS Investment
FAA Reauthorization Act of 2024 (PL 118-63), Section 925	2024	$6M/year (FY2025–FY2028) for UAS test ranges; $13.5M workforce development grants (FY2024)
"Unleashing American Drone Dominance" Executive Order	June 6, 2025	BVLOS rulemaking mandate (30-day deadline); prioritization of U.S.-made drones; export promotion
FAA Part 108 BVLOS NPRM	August 7, 2025	~3,100 public comments; implementation in final regulatory stages
One Big Beautiful Bill Act (OBBBA)	July 4, 2025	$1.4B sUAS industrial base; $250M digital test environment; $2.1B uncrewed surface vessels; $188M maritime robotics R&D

Sources: Public Law 118-63; Executive Order, June 6, 2025; FAA NPRM Docket FAA-2025-1908; OBBBA, July 4, 2025.

A Map of the Money

The federal investment landscape for uncrewed aviation is not a single coherent program. It is a collection of agency investments made from different institutional mandates, on different budget timelines, toward different definitions of success, investments that collectively cover most of the UAS ecosystem but rarely cover it coherently. Understanding where the money comes from, and what each source is designed to accomplish, is a prerequisite for understanding why coordination is so difficult to achieve.

The FAA's investment mandate is, at its core, regulatory infrastructure. The seven federally designated UAS test sites, distributed across the country to represent different geographies, climate conditions, and use-case environments, are the FAA's most visible investment vehicle.[10] The UTM ecosystem represents the FAA's most ambitious infrastructure commitment: a deliberately decentralized architecture in which drone operators, service providers, and the FAA communicate through a distributed network of highly automated systems using application programming interfaces rather than voice communications between pilots and controllers.[4] This design is not accidental. It is what makes the system theoretically scalable to millions of drone operations, a volume that no voice-based air traffic control system could conceivably manage.

NASA's investment role is different in kind from the FAA's and more foundational in nature. The NASA UTM program has been the research engine behind the UTM concept since its inception.[5] NASA does not build commercial products or write operational regulations. It develops the fundamental principles, tests the enabling technologies, and produces the research findings on which the commercial ecosystem and regulatory framework are built. When NASA's FY2026 President's Budget Request for Aeronautics proposed $588.7 million, a 37 percent reduction from the $935 million enacted in FY2025, it was not merely a budget line that changed. Congress rejected the proposed cut; the enacted FY2026 NASA Aeronautics appropriation held at $935 million, the same level as FY2024 and FY2025. The proposed reduction nevertheless signaled a policy priority that the research community will be watching as future budget cycles unfold.[6] It was a compression of the research timeline for UTM and Advanced Air Mobility programs that the entire commercial industry depends upon. When NASA's research budget shrinks, the foundational work slows, and the commercial ecosystem faces a choice: wait for NASA to complete the research on a longer timeline, or fund the research itself through industry consortia and private investment. Neither outcome is efficient.

The Department of Defense (DoD) is the federal government's largest single investor in UAS technology, but its investment mandate is not commercial industry development. DoD invests in drones because drones are increasingly central to military operations, and the gap between domestic manufacturing capacity and military demand has become a national security concern. The Blue UAS Cleared List, maintained by the Defense Innovation Unit (DIU) and refreshed in August 2025 with 23 platforms and 14 framework components, is the government's primary market-shaping tool for domestic drone procurement. The DIU's 2025 Refresh Challenge drew 369 proposals from 19 nations, a measure of the global industry's responsiveness to the U.S. government's demand signal that is, by any standard, remarkable. The Replicator program, launched with urgency to produce 10,000 small drones for operational military use, struggled to translate that urgency into scaled production because it ran headlong into the same chicken-and-egg dilemma that has always challenged defense industrial base investment: manufacturers need guaranteed demand to invest in production capacity, but the government needs production capacity to commit to guaranteed demand.

The Department of Government Efficiency (DOGE)-driven consolidation of Pentagon drone procurement, targeting more than 30,000 small low-cost drones across service branches, represents the latest attempt to break that cycle by creating the demand certainty that commercial manufacturers need to justify capital investment. The lesson of Replicator's production difficulties explicitly informed the design of the OBBBA's $1.4 billion sUAS industrial base investment, a direct appropriation intended to fund the production capacity that the demand signal alone had failed to create.

Beyond the FAA, NASA, and DoD, the federal UAS investment landscape extends into agencies whose primary mandates have nothing to do with aviation but whose programs intersect with drone industry development in consequential ways. The U.S. Department of Agriculture (USDA), with hundreds of open program opportunities, touches drone investment through rural precision agriculture applications, infrastructure inspection programs, and conservation technology grants. The Department of Labor's WIOA framework provides the workforce development funding infrastructure through which community colleges and training providers can build drone technician pipelines. The Economic Development Administration's regional innovation grants support the drone corridor development projects that states and metropolitan areas are pursuing as economic development strategies. None of these agencies has UAS as a primary mandate. None of them, in most cases, coordinates its UAS-adjacent investment with the agencies investing from a different angle. The result is a federal UAS investment ecosystem that is, simultaneously, surprisingly comprehensive and structurally fragmented, covering the industry from almost every direction while rarely speaking in a coherent common voice.

The First Mover Advantage

In the competition among states to capture the economic value of the drone industry, North Dakota is not competing. It has already won the opening round, and the margin of its lead is instructive for every state that is still deciding whether and how to invest.

On February 14, 2025, the North Dakota Senate approved House Bill 1038, an $11 million appropriation to integrate the Northern Plains UAS Test Site's Vantis radar platform with the FAA's Pathfinder radar program.[9] The bill was finalized on February 20. In the compressed timelines of legislative action, this was an unremarkable procedural sequence. In the longer arc of North Dakota's UAS investment strategy, it was the culmination of more than a decade of deliberate, patient, compounding investment in a specific industrial vision.

The $11 million was not speculative capital chasing a trend. It was a precision investment in the final link of an infrastructure chain the state had been building since 2012. North Dakota had already spent more than $100 million in cumulative state investment on UAS infrastructure: the test site, the Vantis ground-based radar network, the University of North Dakota education and research partnerships, the industry recruitment and relationship development that had brought companies including General Atomics, precision agriculture specialists, infrastructure inspection firms, and public safety operators into the state's drone ecosystem. The $11 million was the keystone of an arch that had been under construction for over a decade.

The Vantis-FAA Pathfinder integration expanded North Dakota's BVLOS-capable radar coverage from 3,000 square miles to 56,000 square miles, the largest contiguous BVLOS-capable airspace in the United States.[7] North Dakota became the first state in the nation to integrate with the FAA's Pathfinder program, establishing a federal-state data sharing arrangement that Frank Matus, director at Thales USA and chairman of the North Dakota UAS Council, described as "the first time the federal government had done something like this."

North Dakota poured more than $100 million into UAS infrastructure over a decade, leveraging $400 million in federal and private investment. A four-to-one return ratio that would be exceptional in any investment context.

The financial leverage this investment generates defies easy comparison. The $100 million in cumulative state investment has attracted $400 million in federal and private capital, a four-to-one return ratio that would be exceptional in any investment context. Erin Roesler, then deputy executive director of the Northern Plains UAS Test Site, estimated that by integrating with the Pathfinder program before any other state, North Dakota saves approximately $255 million compared to the cost of later integration, a first-mover advantage measured not in competitive positioning but in actual avoided expenditure.[8]

Former Governor Doug Burgum's characterization of the FAA drone flight data access agreement as a "strategic victory" reflects exactly this dynamic: North Dakota won not because the federal government gave it a gift, but because the state invested strategically over a decade to put itself in the position where the federal opportunity, when it materialized, could be captured quickly and completely. Burgum, who left the governorship in December 2024 to serve as U.S. Secretary of the Interior, built the investment commitment that made that capture possible.

The University of North Dakota's role in this story is not incidental. The UND aviation program has been building UAS curriculum, producing drone-credentialed graduates, and generating research partnerships with the Northern Plains test site for years. When an aerospace company evaluating North Dakota as a potential location for a drone operations center asks about the local talent pipeline, it is not asking about a program that might be built. It is asking about a program that is already producing graduates. That distinction, between potential and demonstrated capacity, is the difference between a state that attracts drone industry investment and one that writes optimistic economic development reports about why it should.

The lesson North Dakota offers is not a template to be photocopied. Its airspace geography, vast, flat, lightly populated, with weather conditions that provide year-round testing variety, is not replicable. Its history as the host of the FAA's first designated UAS test site is not replicable. What is replicable is the strategic logic: make a decision about which industry you intend to lead, commit to the investment timeline that leadership requires, build the infrastructure before you need it, align educational investment with industry development, and exercise the patience to let compound returns accumulate over a decade rather than expecting quarterly results.

When the Money Doesn't Talk to Itself

Return to the airport director. It is now late afternoon on the same Monday, and she has spent most of the day reading grant applications and making notes. She has not made any decisions, because every program she is looking at requires information she does not yet have: program-specific data formats, letters of support from partner organizations that are themselves overwhelmed, budget templates that differ from each other in ways that are just different enough to prevent any economies of scale in preparing them. She has identified four programs that could collectively fund something close to her airport's UAS vision. She has also identified that pursuing all four simultaneously would consume essentially the entire discretionary capacity of her small administrative team for the better part of a year.

This is the face of federal fragmentation, not malice, not incompetence, but structural misalignment between the way the federal government organizes its investment and the way local institutions must organize their operations. The FAA, NASA, EDA, DOL, USDA, and DoD all have genuine interest in building the UAS ecosystem. They have programs, funding, staff, and program officers dedicated to different pieces of the problem. What they do not have is any mechanism for talking to each other about it, for aligning their grant cycles, harmonizing their definitions of eligible activities, sharing data about their grantees, or ensuring that a WIOA-funded drone technician training program at a community college is automatically connected to the FAA UAS Collegiate Training Initiative (CTI) network and the EDA-funded regional drone corridor development project operating in the same geographic area.

The cost of this fragmentation is not theoretical. It is measured in the administrative hours that regional airport directors, community college workforce deans, and economic development officials spend writing and managing multiple grant applications instead of running the programs those applications are designed to fund. It is measured in the duplicated reporting requirements that consume the institutional capacity that should be directed toward training students and building industry partnerships. It is measured in the investment opportunities that capable local institutions miss because no one in the federal system has told them that a program relevant to their strategic priorities even exists. And it is measured, ultimately, in the workforce that does not get trained, the infrastructure that does not get built, and the regional economic development that does not happen, not because the money was unavailable but because the connective tissue between the money and the local actors who need it was absent.

The fragmentation problem is not new, and it is not unique to uncrewed aviation. What makes it particularly costly in this sector is the window of competitive opportunity that the 2025 legislative surge has opened. Fragmented federal investment in a slow-moving industry is inefficient. Fragmented federal investment in a rapidly-moving industry is costly in a more urgent way: the window during which coordination would compound returns is precisely the window during which fragmentation is most expensive.

The Strategy We Don't Yet Have

The federal government has, on at least two occasions in recent history, demonstrated that it knows how to coordinate industrial investment at scale. The Interstate Highway System required a federal coordinating architecture that aligned state highway departments, federal construction standards, land acquisition authorities, and long-term funding commitments into a coherent national infrastructure strategy. The Creating Helpful Incentives to Produce Semiconductors (CHIPS) and Science Act of 2022 created the CHIPS Program Office within the Department of Commerce, a dedicated investment vehicle with a clear mandate, cross-agency coordination authority, and the ability to align public investment with private capital through a structured incentive framework that made private semiconductor investment commercially rational in locations where it previously would not have been. In both cases, the key insight was the same: federal investment achieves its strategic purpose not when it is simply appropriated, but when it is organized, when the money flows through an architecture deliberately designed to compound its effects.

The CHIPS Act is the most instructive model for uncrewed aviation, not because semiconductors and drones are identical technologies, but because the CHIPS Act solved exactly the coordination problem that UAS investment currently faces. The semiconductor industry's supply chain dependence on foreign manufacturing had become a national security vulnerability that required not just investment but organized investment: a mechanism for ensuring that federal dollars, private capital, workforce development resources, and research infrastructure arrived in the same place at the same time in a sequence designed to produce working semiconductor fabs rather than disconnected grants. The CHIPS Program Office provided that mechanism. The result: hundreds of billions of dollars in private semiconductor investment flowing to American manufacturing facilities that would not have been built without the federal coordination architecture that made the investment commercially viable.

A UAS equivalent would not require the creation of a new agency. It would require a coordinating function, an interagency task force or office, perhaps housed within the Office of Science and Technology Policy or the National Economic Council, with the explicit mandate and authority to align the FAA's regulatory investment, NASA's foundational research, DoD's procurement commitments, DOL's workforce development funding, USDA's rural application programs, and EDA's regional economic development resources around a shared geographic and industrial strategy. The FAA's seven designated test sites provide a natural organizing geography for exactly this kind of aligned investment: each test site state could serve as the anchor for a regional UAS investment cluster, with federal workforce development, research, manufacturing incentive, and commercial deployment investments coordinated around a shared set of outcomes rather than administered through separate agency programs with separate metrics and separate program officers.

What 2025's extraordinary legislative and executive investment created was the resources for a national UAS strategy. What remains to be built is the architecture.

The competitive urgency of this challenge should not be understated. China's investment in its drone industry, through state-directed capital, subsidized manufacturing, integrated civil-military development, and an export strategy that has made DJI the world's dominant commercial drone brand, represents the external pressure that the "Unleashing American Drone Dominance" executive order was written to address. The domestic response to that pressure cannot succeed if it remains fragmented. American democratic governance does not coordinate investment the way China's state-directed system does, and that difference is real. But American federalism has demonstrated, repeatedly, that it can create the coordinating mechanisms, the program offices, the interagency task forces, the grant frameworks, that allow federal investment to function as coherent industrial strategy rather than as a collection of well-intentioned programs that do not talk to each other.

The money is real. Summer 2025 proved that Congress and the executive branch can mobilize serious capital for uncrewed aviation. North Dakota proved that patient, coordinated, leveraged public investment can produce extraordinary results. The airport director's overwhelmed inbox proved that the architecture for deploying that capital in a way that reaches and empowers the local actors who will actually build the industry remains incomplete. What the United States has built is the financial foundation of a national UAS strategy. What it has not yet built is the strategy itself, the connective tissue, the coordinating mechanisms, the aligned incentive frameworks that would allow the billions already committed to produce the national leadership in uncrewed aviation that the executive orders have promised and the industry requires. That architecture is the work still to be done. And it is, in many respects, more difficult than writing the check.

The money is real. The intent is genuine. The architecture that would allow it to compound into national industrial leadership remains to be built.

Looking Ahead — Chapter Ten

Yet even the most ambitious coordinated public investment is not the primary capital story in uncrewed aviation. The private capital flowing into the drone startup ecosystem has dwarfed federal spending in volume, velocity, and geographic reach. Understanding where that capital is going, which technologies it favors, which geographies it concentrates in, which founders it reaches and which it consistently passes over, is essential to understanding how the uncrewed aviation industry will actually develop over the next decade. That is the story Chapter Ten tells.

Notes and Sources

1. "Unleashing American Drone Dominance." Executive Order. The White House, June 6, 2025. Directs FAA to issue BVLOS proposed rulemaking within 30 days; prioritizes procurement of U.S.-manufactured UAS by federal agencies; calls for streamlining of certification processes; promotes export of American UAS technology.

2. One Big Beautiful Bill Act (OBBBA). Signed July 4, 2025. UAS and defense provisions include $1.4 billion to scale the sUAS industrial base; $250 million for a digital test environment for autonomy validation; $2.1 billion for medium uncrewed surface vessel development; $188 million for maritime robotics research and development.

3. FAA Reauthorization Act of 2024. Public Law 118-63, Section 925, enacted May 16, 2024. Establishes $6 million per year for fiscal years 2025 through 2028 for FAA-designated UAS test sites, structured as matching funds for commercial technology demonstration and validation partnerships.

4. FAA. "Unmanned Aircraft System Traffic Management (UTM)." FAA.gov, accessed May 2026. Describes the UTM ecosystem architecture, Near-Term Approval Process (NTAP) for service providers, and the FAA's ongoing collaboration with NASA and industry to enable scalable low-altitude airspace management.

5. NASA. "UTM: UAS Traffic Management." NTIS/NASA Publication N240013848; presented at NASA/Bay Area Public Operators UAS Engagement, Mountain View, CA, October 29, 2024; National Technical Reports Library. Covers collaborative airspace management principles, service provider frameworks, and key site operations for enabling BVLOS operations.

6. NASA FY2026 Aeronautics Budget. President's Budget Request: $588.7 million, a proposed 37 percent reduction from the $935 million enacted in FY2025. Enacted FY2026 appropriation (P.L. 119-74, signed January 23, 2026): $935 million, identical to FY2024 and FY2025 enacted levels. Congress rejected the Administration's proposed cut. Sources: U.S. Senate Committee on Commerce, Science, and Transportation press release, January 2026; Congressional Research Service, NASA Appropriations and Authorizations: At a Glance (R43419), updated February 27, 2026; SpacePolicyOnline.com, January 2026.

7. Michael Standaert. "Working with FAA expands North Dakota commercial drone prospects." North Dakota Monitor, February 28, 2025. Reports on House Bill 1038 and the planned integration of the Northern Plains UAS Test Site's Vantis radar platform with the FAA's Pathfinder program, including estimated coverage expansion and industry implications.

8. Emma Larson. "North Dakota Just Became the Drone Capital with FAA Data." KVRR / North Dakota Real Estate & Business, April 19, 2026. Reports on Former Governor Doug Burgum's characterization of the FAA drone flight data access agreement as a "strategic victory"; includes estimates of R&D acceleration and development cost savings for companies operating in the state. Note: Burgum left the North Dakota governorship in December 2024 and serves as U.S. Secretary of the Interior.

9. North Dakota Legislature. House Bill 1038 — UAS Infrastructure Appropriation. Signed February 2025. Approves $11 million for integration of Northern Plains UAS Test Site Vantis radar platform with the FAA's Pathfinder radar program, expanding BVLOS-capable radar coverage from 3,000 to 56,000 square miles.

10. FAA. "UAS Broad Agency Announcement Program." FAA.gov, updated March 30, 2026. Describes the FAA's research and technology investment framework for uncrewed aviation, including partnerships with industry and academic institutions for technology demonstration and validation at designated test sites.

11. FAA. "Funding Opportunities — Aviation Workforce Development Grants." FAA.gov, updated January 8, 2026. Documents $13.5 million awarded in 22 grants in fiscal year 2024, explicitly authorized under the FAA Reauthorization Act for recruitment of UAS operators and aviation maintenance workers through education, outreach, and community partnerships.

12. American Security Drone Act (ASDA). Enacted as part of the National Defense Authorization Act for Fiscal Year 2024, December 22, 2023; FAR interim rule implementing ASDA procurement restrictions effective November 12, 2024. Establishes the statutory basis for exclusion of covered foreign-manufactured UAS from federal procurement, creating the domestic demand signal that OBBBA industrial base investments are designed to answer.

Private Capital and the Uncrewed Aviation Ecosystem

"Patient capital and proven contracts build industries. Impatient capital and ambitious decks build cautionary tales. The drone sector has paid for both lessons, and the bill was expensive enough to matter."

— *Sharon Rossmark*

The Pitch That Changed

In the spring of 2021, a drone startup founder walked into a conference room in San Francisco carrying a pitch deck, a prototype video, and the ambient confidence of an entrepreneur who had read the room correctly. The room, that season, was suffused with a particular kind of enthusiasm, the gold-rush electricity of a technology investment cycle that had not yet been required to justify itself in quarterly earnings calls or deliver units on a production schedule. The investors across the table were excited. The questions were general: market size, regulatory tailwinds, team pedigree. The term sheet, when it came, came quickly. The valuation it implied assumed that she would be operating at meaningful commercial scale within eighteen months. No one in the room asked about unit economics. No one asked about Department of Defense (DoD) contract specifics. Manufacturing was described, in the abstract, as something to be figured out later. Later, in that spring, felt very close.

Four years passed. In early 2025, the same founder walked into a different kind of room. The investors were still there, two traditional venture funds that had survived the intervening correction with their conviction intact. But alongside them sat a representative from the corporate venture arm of a defense prime and a principal from a sovereign wealth fund whose mandate included critical technology infrastructure. The questions were different in kind, not merely in degree. They asked for unit economics broken out by commercial and defense revenue streams. They asked which DoD programs her technology was qualified under and what the re-qualification pathway looked like for the specific Air Force contract she was pursuing. They asked for her domestic component sourcing plan, her National Defense Authorization Act (NDAA) compliance certification timeline, and her manufacturing capacity at twelve months and thirty-six months from a Series B close.

The drone investment market did not disappear between 2021 and 2025. It matured, painfully, necessarily, and in some ways productively. The $3.67 billion that flowed into the sector at the 2021 peak taught the industry, and its investors, what drone businesses actually look like when they have to perform rather than promise.[1] Some of that learning came at enormous cost. Lilium Jet's collapse into insolvency is one invoice. The evaporation of paper valuations across a cohort of Special Purpose Acquisition Company (SPAC)-listed electric vertical takeoff and landing (eVTOL) companies is another. But the market that emerged from the correction, more disciplined, more strategically anchored in defense and enterprise demand rather than consumer enthusiasm, more focused on the demonstrated contract pipeline rather than the projected adoption curve, is, in many ways, better positioned to produce durable companies than the one that preceded it.

The drone investment market did not disappear between 2021 and 2025. It matured, painfully, necessarily, and in some ways productively.

Private capital has been the primary engine of the drone startup ecosystem since the sector's earliest days, and understanding its flows, where it has gone, where it has conspicuously not gone, and where it is heading in the legislative and operational environment that took shape in the summer of 2025, is essential to understanding which companies will have the resources to define the industry's next decade. The capital story is not a simple narrative of boom and bust. It is a story of a market learning, through the most expensive curriculum available, what the drone business actually requires.

What the Hype Cycle Cost, and What It Built

Begin with the numbers, because the numbers establish the scale of both the opportunity and the correction. Global drone industry funding peaked at $3.67 billion in 2021, the high-water mark of a technology investment supercycle that simultaneously inflated valuations across electric vehicles, fintech, and consumer software. By 2024, that figure had declined to $820 million, a 73 percent reduction from the peak.[1] The trajectory follows the Gartner Hype Cycle with a precision that the model's inventors might have found embarrassing: a technology trigger in the mid-2010s, as commercial drone regulations began to take shape; a peak of inflated expectations in 2020 and 2021, as pandemic-era demand for contactless logistics and remote inspection collided with the lowest interest rate environment in modern financial history; and then the trough of disillusionment that 2022 through 2024 made visible, as companies that had gone public at valuations requiring rapid regulatory approval and swift manufacturing scale found neither on the timeline investors had expected.

The correction was not a collapse. Average funding per round, even in the contracted 2024 market, held at $27.2 million, indicating that investors did not abandon the sector but rather became far more selective within it. The trough of disillusionment in the Gartner model is not a dead end; it is the phase in which only companies with viable business models attract sustained investment, and the companies that survive it are positioned for the slope of enlightenment that follows.

No episode in the uncrewed aviation investment story illustrates both the potential and the peril of the 2020–2022 era more completely than the rise and fall of Lilium. The eVTOL sector and the commercial drone sector drew from the same investor community and were shaped by the same capital cycle dynamics — the lessons of one apply directly to understanding the other. Lilium was a genuine technology company by any serious measure: it raised more than $1 billion from investors including Tencent, went public via SPAC merger with Qell in 2021 on the Nasdaq, and secured high-profile orders including 100 electric jets from Saudi Arabia. Its engineering team was building hardware of extraordinary complexity, a novel ducted-fan eVTOL architecture unlike anything that had been certified for commercial aviation. The technology was real, and the ambition behind it was legitimate. What Lilium could not survive was the mismatch between the timeline required to develop and certify a revolutionary aircraft and the timeline that SPAC investors' liquidity expectations imposed on the company from the moment it went public.

The SPAC structure provided access to public market capital without the traditional IPO underwriting process, a genuine advantage for capital-intensive deep-tech companies. What it could not provide was a public market patient enough to wait for deep-tech development timelines. The mismatch was not a flaw in the technology. It was a structural incompatibility between the liquidity expectations of SPAC investors and the time horizons that aircraft development, FAA certification, and manufacturing scale-up actually require.

Going public via SPAC provided capital access, but it simultaneously imposed quarterly reporting requirements, earnings call scrutiny, and investor pressure calibrated to the cadence of public markets on a company that needed years, not quarters, to achieve the milestones that justified its valuation. The cash ran out before the product could fly commercially. In October 2024, Lilium filed for insolvency. Archer Aviation won the bankruptcy bid process: 300 patents spanning high-voltage systems, flight controls, ducted fans, and advanced aircraft design, acquired for €18 million, approximately $21 million, a price that, measured against the more than $1 billion Lilium had consumed, illustrated with painful clarity that the intellectual property of a failed company can long outlast the company itself.[4]

The contrast with Archer Aviation illuminates what survival in the SPAC era actually required. Archer, too, went public via SPAC in 2021, and it carried into the public markets the same set of expectations that ultimately broke Lilium. But Archer's trajectory after the correction is a story of strategic pivots that gave the company multiple paths to revenue while its core certification timeline extended. By 2025 and into 2026, Archer had exceeded 400 test flights, built a total portfolio exceeding 1,000 patents, including the 300 acquired from Lilium's estate, and secured United Airlines and Stellantis as strategic investors, with Stellantis committing manufacturing labor and capital expenditures representing up to $400 million in additional capital access toward scaling the Midnight aircraft.[6] Archer added an Anduril defense partnership, extending its addressable market into military applications, and was selected as the official air taxi provider for the 2028 Los Angeles Olympics. Archer's total funding has surpassed $1.5 billion.[5]

Table 10.1 — Global Drone Industry Funding Trajectory, 2021–2026

Year	Global Funding	Key Development
2021	$3.67 billion (peak)	SPAC boom; Joby, Archer, Lilium go public; peak inflated expectations
2022	Significant decline begins	Interest rate environment shifts; SPAC valuations compress; trough of disillusionment begins
2023	Contraction continues	VC-backed defense startups raise $2.7 billion; defense-aligned companies counter-cycle
2024	$820 million	73% decline from peak; Lilium insolvency; avg. round still $27.2M; defense VC rises to $3B
2025–2026	Selective recovery, strategic capital	OBBBA $1.4B; Shield AI $2B+ Series G; Skydio SkyForge $3.5B; new investment thesis emerges

Sources: Drone Industry Insights; StartUs Insights (July 2025); Caerobotics / Saxena (July 2025); Crunchbase; company announcements.

From Hype to Thesis

If there is a single company whose funding trajectory tells the story of the new investment reality more clearly than any other, it is Shield AI. In March 2025, Shield AI raised $240 million in a Series F round at a $5.3 billion valuation, notable primarily for the identity of its lead investors.[2] L3Harris Technologies, one of the largest traditional defense primes in the United States, and Hanwha Aerospace, South Korea's leading defense and aerospace manufacturer, co-led the round. Andreessen Horowitz participated alongside them. These were not generalist technology investors deploying capital against an adoption curve. They were strategic investors, companies with operational reasons to want the technology to succeed, and the institutional patience to stay in through development timelines that had broken SPAC investors.

On March 26, 2026, Shield AI announced a Series G that closed at $2 billion in new equity, establishing a post-money valuation of $12.7 billion, a 140 percent increase in twelve months. Advent International led the equity tranche. JPMorgan Chase's Security and Resiliency Initiative co-led. Blackstone provided $500 million in fixed-return preferred equity. An additional $250 million delayed draw facility brought the total capital event to approximately $2.25 billion. Shield AI's projected 2026 revenue stands at $540 million or more.[3]

What made Shield AI's Series G credible at this scale was not a pitch deck. It was a contract.

What made the Series G credible at this scale was not a pitch deck. It was a contract: the United States Air Force's selection of Shield AI's Hivemind autonomy platform as the software provider for the Collaborative Combat Aircraft (CCA) program, autonomous wingmen designed to fly alongside crewed F-35s and the forthcoming F-47. With a production backlog tied to one of the most consequential autonomous systems programs in Air Force history, Shield AI is no longer a bet on future technology. It is a bet on a company that has already won. That distinction, between a company with a compelling technology thesis and a company with a demonstrated contract pipeline, is the defining characteristic of what the post-correction investment market rewards.

The broader 2025–2026 investment landscape produced a cohort of well-capitalized companies across multiple unmanned aircraft systems (UAS) segments, each reflecting a different dimension of the new investment thesis. Skydio's $110 million Series F, closed at a $4.4 billion valuation in April 2026, was announced alongside the company's $3.5 billion SkyForge manufacturing expansion program, projecting 5,000 or more jobs and validating the domestic manufacturing thesis that federal procurement policy had been building toward.[8] Quantum Systems closed a €160 million Series C in May 2025 to scale dual-use autonomous aircraft technologies. Firestorm Labs raised $47 million in a Series A for modular domestic drone production. Saronic Technologies raised $600 million in a Series C at a $4 billion valuation for autonomous surface vessels.[7] Epirus raised $250 million in a Series D for counter-drone and directed energy systems. The AIRO Group completed a public offering in June 2025. Unusual Machines saw its stock surge 40 percent following Pentagon procurement budget announcements.[9] Red Cat Holdings (Nasdaq: RCAT), whose Teal Drones subsidiary holds the U.S. Army's Short Range

Reconnaissance Program of Record contract, closed a $46.75 million registered direct offering in June 2025, demonstrating precisely the investment dynamic the post-correction market rewards: a domestic, NDAA-compliant drone manufacturer with a verified defense contract pipeline attracting institutional capital on the strength of operational proof rather than projected adoption curves.[13]

VC-backed defense startups across the broader sector raised $3 billion in 2024, up 11 percent from $2.7 billion in 2023, a counter-cyclical signal that defense-aligned drone and autonomy companies continued to attract capital even as the commercial drone market contracted.[2] When the legislative surge of summer 2025 arrived, the One Big Beautiful Bill Act's (OBBBA) $1.4 billion for the small UAS industrial base, the Unleashing American Drone Dominance executive order, the DoD procurement directive, it catalyzed a fresh wave of private investment, with hundreds of millions in new funding announced in the May through July 2025 window alone.[9] The correction of 2022–2024 did not kill the drone investment market. It pruned it, leaving the companies with genuine business models, genuine contracts, and genuine technology better capitalized and better positioned than the crowded, thesis-light field of 2021.

When the Smart Money Has an Operational Reason

The most significant structural shift in drone and autonomy investment is not its volume. It is its composition. The investors leading the largest rounds in 2025 and 2026 are not generalist venture funds deploying capital against a thesis about technology adoption curves. They are companies, financial institutions, and sovereign actors that have operational reasons to want the technology to succeed, and the balance sheet resources to stay in through the development timelines that proved fatal to SPAC investors. This shift in capital composition represents a fundamental improvement in the quality of investment the industry is receiving, even as the headline dollar volumes remain far below the 2021 peak.

JPMorgan Chase's Security and Resiliency Initiative represents perhaps the most consequential new entrant to the space. When the largest bank in the United States creates a $10 billion commitment specifically structured to back U.S. and allied defense, energy, and critical-mineral firms and then co-leads a $2 billion equity round for the company whose autonomy software is powering the Air Force's next-generation combat aircraft program, it is doing something more significant than making a financial bet.[10] It is signaling where the financial system's center of gravity is moving. The autonomous systems sector is no longer a technology niche attracting specialized technology investors. It is infrastructure, and infrastructure, as any banker with a long memory knows, attracts a different class of capital entirely: patient, strategic, and institutionally anchored in a way that venture fund quarterly marks cannot replicate.

Hanwha Aerospace's participation in Shield AI's Series F reflects a different but equally important logic: allied-nation defense primes are investing directly in U.S. autonomy startups because the technology being developed will be deployed in allied-nation military systems, not merely American ones. The DIU's 2025 Refresh Challenge drew proposals from 19 nations. Shield AI delivers Hivemind-powered systems to 29 allied nations. The drone and autonomy market is, in its most consequential applications, a global defense market. The strategic investors positioning themselves within it are thinking globally even when they are writing checks to American companies.

Table 10.2 — Strategic Investor Snapshot, 2025–2026

Investor	Role and Significance
JPMorgan Chase Security and Resiliency Initiative	$10B commitment to U.S. and allied defense, energy, and critical-mineral firms; co-led Shield AI $2B Series G. Wall Street's largest bank formally enters autonomous systems.
L3Harris Technologies	Co-led Shield AI $240M Series F. A traditional defense prime investing in the autonomy startup whose software could replace legacy ISR platforms.
Hanwha Aerospace (South Korea)	Co-led Shield AI Series F. An allied-nation prime making a direct bet on U.S. autonomous capability for allied deployment.
Stellantis	Strategic investor in Archer; manufacturing labor and capital worth up to $400M. The automotive sector's bet on eVTOL as its next industrial platform.
United Airlines	Strategic investor in Archer. An airline betting on urban air mobility as its next growth adjacency and feeder network.
Andreessen Horowitz	Consistent across Shield AI Series D, E, and F. Traditional VC's deepest and most sustained commitment to the autonomy sector.

Sources: Crunchbase; company announcements; cited notes throughout this chapter.

Stellantis's commitment to Archer illustrates the strategic investor logic from the industrial rather than defense angle. The automotive prime is not investing in Archer because it has concluded that urban air mobility will replace ground transportation. It is investing because it sees Archer's manufacturing challenge, precision assembly of a complex electromechanical aircraft at production volumes that commercial air mobility requires, as a problem its industrial capabilities are uniquely positioned to solve. The same engineering culture that optimizes battery management systems, produces precision automotive components at scale, and manages the complex supplier logistics of modern vehicle manufacturing is directly applicable to aircraft production. Stellantis is not merely a financial investor in Archer. It is acquiring a seat inside the eVTOL supply chain before the market matures, and it is doing so by contributing its most valuable asset: its manufacturing knowledge base.

The Investment Blind Spots

There is a version of the drone investment story that is, by the standards of venture capital, a success narrative. The companies that survived the correction are better capitalized, better contracted, and better positioned for sustainable growth than they have ever been. Shield AI's $12.7 billion valuation and Skydio's $3.5 billion manufacturing commitment are not small achievements. They represent real technological progress, genuine contract wins, and investor confidence earned rather than assumed. But an honest account of where private capital has flowed in the drone and autonomy sector requires equal attention to where it has not gone, and why those absences represent both market failures and strategic risks for the industry's long-term health.

The workforce infrastructure blind spot is the most consequential, and it is almost entirely invisible in the press release record of drone investment activity. Across the billions of dollars committed to autonomy platforms, manufacturing scale-up, and defense contract pursuit in 2025 and 2026, there is virtually no explicit private investment in the workforce that will build, operate, and maintain what the capital creates. The community college running a UAS Collegiate Training Initiative (UAS-CTI) program. The Unmanned Safety Institute (USI) launching twelve-week maintenance technician programs. The organizations building the visibility and community infrastructure that recruits the next generation of drone professionals from communities that the industry has not yet seriously engaged. None of these attract venture capital, because none of them are venture-scale businesses. They require grant funding, philanthropic investment, and government procurement to operate, and they are, in a fundamental sense, the most important investment the industry can make.

No amount of autonomy software or domestic manufacturing capacity produces output without the humans who can deploy, maintain, and operate it.

The domestic component manufacturing gap is the second major blind spot. The battery cells, the rare earth-derived sensors and motors, the permanent magnets that power drone propulsion systems, the supply chain that remains China's most powerful tool for economic leverage over the U.S. drone industry, are largely unaddressed by private capital at the scale required to matter. Skydio's $3.5 billion SkyForge commitment, directed specifically at domestic component suppliers as well as drone assembly, is the most serious private attempt to build domestic supply chain sovereignty at industrial scale. But it represents one company's strategy and one manufacturing platform's footprint, not an industry-wide investment pattern. The venture capital market is not structurally suited to fund the patient, capital-intensive, low-multiple manufacturing investments required to rebuild component supply chains that China has spent decades and trillions developing. That gap requires a class of capital, industrial development finance, patient equity with long hold periods, government co-investment, that the current drone investment infrastructure has not yet organized to deliver.

The mid-market commercial drone services gap receives the least attention, and that invisibility is itself part of the problem. Between the venture-backed startups raising nine-figure rounds and the bootstrapped solo operators building small Part 107 service businesses, there exists a large and largely overlooked population of commercial drone services companies, infrastructure inspection firms, precision agricultural operators, enterprise UAS service providers, building real businesses at $5 to $50 million in annual revenue. These companies constitute, in aggregate, the bulk of the commercial drone industry's actual economic activity. They employ the most people. They generate the most flight hours. They produce the most data about real-world UAS utility in real commercial applications. And they are poorly served by the investment infrastructure that has formed around the high-profile startup sector, too small for institutional venture capital, whose minimum viable check sizes have risen as the sector has consolidated, and too capital-intensive for bootstrapping.

What the Next Wave Looks Like

The investors positioning most intelligently in the drone and autonomy market share a common and consequential insight: the regulatory clearing of the BVLOS era will be the single most significant demand catalyst the commercial drone industry has ever experienced. The FAA's Part 108 BVLOS framework, published in August 2025, is in its final regulatory stages.[12] When that framework is operational, the commercial applications that currently require waiver-by-waiver airspace authorization, infrastructure inspection across linear corridors, precision agricultural services across large tracts, last-mile logistics in complex urban environments, will become routine, scalable, and insurable. The companies that are capitalized, operational, and proven ahead of that regulatory clearing will capture the growth wave that follows.

Five investment theses are drawing the most sophisticated capital, each mapping onto a different dimension of the BVLOS opportunity. Autonomy software platforms, particularly those capable of being licensed across multiple hardware form factors, represent the highest-multiple business model in the sector. Counter-UAS and drone security systems represent a permanent and growing market: Epirus's $250 million Series D reflects investor conviction that the proliferation of drones has created a security demand that will not diminish regardless of what the commercial drone market does in any given quarter. Domestic NDAA-compliant manufacturing, in which Skydio's SkyForge commitment is the template and the standard, represents the supply-side response to the demand signal that federal procurement policy has been building since the original NDAA Section 848 restrictions. BVLOS enablement infrastructure, the detect-and-avoid systems, UTM integration platforms, and sensor suites that make BVLOS commercially viable, represents the picks-and-shovels investment in the BVLOS opportunity: the enabling layer that every BVLOS operator will need, regardless of which airframes or mission profiles win in their specific market. And dual-use platforms, systems capable of serving DoD, commercial infrastructure operators, and allied-nation governments through a single core technology investment, command the highest valuations in the sector because they carry multiple revenue paths through the same research and development expenditure.

What is notably, and consequentially, absent from this list is the human capital infrastructure that makes all of the above viable at scale. The workforce training programs that produce licensed UAS operators and qualified maintenance technicians. The credential platforms that create portable professional identities for the growing UAS workforce. The community-building organizations that recruit talent from communities the industry has not yet seriously reached, communities whose engagement is not merely a strategic preference but a practical necessity for an industry that will need tens of thousands of trained professionals in the next decade to operate what the capital is building. These are not traditional investment opportunities, and they will not appear in the funding announcements that define the public investment narrative. The gap between the capital flowing into the supply side of the drone and autonomy sector and the capital available for the human capital infrastructure that the sector requires is, in many ways, the most important structural challenge the industry faces in the years immediately ahead.

Looking Ahead — Chapter Eleven

Closing those gaps requires understanding the role of the largest single investor in the uncrewed aviation ecosystem: the United States Department of Defense, whose procurement decisions, research and development investments, and operational requirements do not merely complement the private capital landscape. They define its boundaries, its incentives, and its opportunities. That story is the subject of Chapter Eleven.

Notes and Sources

1. Investment funding figures (global drone industry peak of $3.67 billion in 2021; $820 million in 2024; 73% decline from peak): Drone Industry Insights (droneii.com), the primary global source for commercial drone investment data, annual funding reports 2021–2024, corroborated by Caerobotics / Dr. Ruchi Saxena, "Global Drone Sector Investments in 2025," Caerobotics.com, July 10, 2025. Average funding per round ($27.2 million), investor count (2,000+), and

funding round count (7,000+): StartUs Insights, Drone Industry Report 2025, startus-insights.com, July 2025.

2. Chris Metinko. "Defense Tech Startup Shield AI Hits $5.3B Valuation Mark." Crunchbase News, March 6, 2025. Shield AI $240 million Series F; L3Harris and Hanwha Aerospace as lead strategic investors; Andreessen Horowitz participation; $3 billion in VC-backed defense startup funding in 2024, up 11% from $2.7 billion in 2023.

3. Nadia Dubois. "Shield AI's $2B Series G at $12.7B: Inside the CCA Win." May 5, 2026. Shield AI $2 billion Series G; $12.7 billion post-money valuation; Advent International as lead; JPMorgan Chase Security and Resiliency Initiative as co-lead; Blackstone $500 million preferred equity tranche; $250 million delayed draw facility; projected 2026 revenue $540 million+; U.S. Air Force CCA program selection.

4. Kirsten Korosec. "Defunct electric aircraft startup Lilium's tech lives on over at Archer." TechCrunch, October 16, 2025. Lilium insolvency filing; competitive bid process; Archer acquisition of 300 Lilium patents for €18 million ($21 million); Joby Aviation and Ambitious Air Mobility Group as competing bidders.

5. Sterne Kessler. "Turbulence in the eVTOL IP Wars – Joby Attempts to Ground Archer." November 21, 2025. Joby Aviation v. Archer trade secrets litigation; Archer funding exceeding $1.5 billion; 1,000+ patent portfolio; LA 2028 Olympics air taxi selection.

6. Mike Hirschberg. "eVTOL Leaders Find Uneven Success." Vertiflite (eVTOL News), November/December 2024. Archer additional capital raises; Stellantis manufacturing labor and capital commitment (up to $400 million); 400+ Midnight test flights; United Airlines strategic investment.

7. Dom Guzman. "Saronic Technologies $600M Series C at $4 Billion Valuation." Crunchbase News, 2025. Autonomous surface vessels; Elad Gil as lead investor.

8. Skydio, Inc. SkyForge Program Announcement. April 2026. $3.5 billion domestic manufacturing expansion; $110 million Series F at $4.4 billion valuation; 5,000+ projected jobs; 60,000+ drones delivered; all U.S. military branches and 29 allied nations as customers.

9. Dronelife / InnovateEnergy. "How New U.S. Laws Are Unleashing a Wave of Investment in American Drones." August 2025. Post-executive order and OBBBA investment surge; Firestorm Labs $47 million Series A; Quantum Systems €160 million Series C; AIRO Group IPO June 2025; Unusual Machines 40% stock surge on Pentagon procurement announcements.

10. JPMorgan Chase. Security and Resiliency Initiative. 2025. $10 billion commitment to U.S. and allied defense, energy, and critical-mineral firms; participation as co-lead in Shield AI Series G.

11. One Big Beautiful Bill Act (OBBBA). Signed July 4, 2025. Section providing $1.4 billion for the small Unmanned Aircraft System industrial base.

12. Federal Aviation Administration. Notice of Proposed Rulemaking: Operation of Unmanned Aircraft Systems Beyond Visual Line of Sight (Part 108). Published August 7, 2025. Implementation in final regulatory stages.

13. Red Cat Holdings, Inc. (Nasdaq: RCAT). Registered direct offering announcement, June 18, 2025. Gross proceeds $46.75 million. Teal Drones subsidiary holds U.S. Army Short Range Reconnaissance (SRR) Program of Record contract. Company operates through Teal Drones and FlightWave Aerospace subsidiaries; products include Black Widow ISR system, TRICHON fixed-wing VTOL, and FANG NDAA-compliant FPV drones. Source: Globe Newswire / Red Cat Holdings press release, June 18, 2025.

Defense Dollars and Dual-Use Technology

"The history of American technological leadership is, at its foundation, a history of public investment in capabilities whose commercial applications were not yet imagined. Defense does not merely buy weapons. It builds industries."

— Sharon Rossmark

The Hangar That Explains Everything

The venue is a former aircraft hangar, and the floor is arranged like a science fair, if the science fair had a $2 billion procurement budget and a room full of military acquisition officers asking questions that no hobbyist drone enthusiast would recognize. This is a Defense Innovation Unit (DIU) pitch day: dozens of drone and autonomy startups have set up demonstration stations, and the officers moving methodically from table to table are not browsing. They are evaluating. The questions are operational: What is your current technology readiness level? What is your domestic component sourcing plan? What does your NDAA compliance certification look like? How long to deploy at scale from a Series B close? This is not a networking event. It is a market. And the fact that the Department of Defense (DoD) is running it, reaching out to commercial innovators rather than waiting for them to navigate the traditional acquisition labyrinth, tells you more about the current state of uncrewed systems than any Pentagon press release.

The scene is not an outlier. It is the new normal. Over the past decade, the relationship between the Department of Defense and the commercial drone industry has undergone a structural transformation that neither side fully anticipated. The DoD, historically the originating source of aviation technology, has found itself in the unfamiliar position of purchasing commercial innovation, pulling autonomous navigation software, computer vision algorithms, and rapid manufacturing techniques from the startup ecosystem because those capabilities are advancing faster in the commercial sector than traditional defense acquisition cycles can develop them internally. The commercial industry, meanwhile, has discovered that the

DoD is not merely a customer but a platform: a source of technology validation, procurement signal, workforce talent, and market-shaping standards whose influence extends far beyond the defense contracts themselves.

This chapter maps that relationship, its mechanisms, its tensions, and its implications for every actor in the uncrewed aviation ecosystem. Understanding how defense dollars flow into and through the commercial drone market is not optional background for anyone building, investing in, or regulating this industry. It is essential foreground.

The Most Powerful Customer in the Drone Economy

No actor in the U.S. uncrewed aviation ecosystem exercises more influence over the industry's direction than the Department of Defense. Across the Army, Navy, Air Force, Marine Corps, and Special Operations Command, annual UAS procurement and research and development spending runs into the tens of billions of dollars, a figure that dwarfs the combined venture capital flowing into the commercial drone sector in any given year. These dollars do more than purchase aircraft. They set the technological agenda, shape the manufacturing base, define the trust standards, and determine which companies have the contract pipeline required to attract the institutional capital that funds the next generation of platforms.

DoD's influence operates through three primary channels. The first is technology validation. When the DoD buys a capability and deploys it at operational scale, it signals to the rest of the market, to commercial buyers, to institutional investors, to domestic manufacturers evaluating whether to retool for drone production, that the technology in question is mature enough for environments where failure has real consequences. Shield AI's selection as the autonomy software provider for the Air Force's Collaborative Combat Aircraft (CCA) program did not merely give Shield AI a contract. It gave every investor evaluating the autonomous systems sector a reference point that no pitch deck could provide.[10]

The second channel is the establishment of security standards. The Blue UAS Cleared List, administered by the DIU and updated most recently in August 2025 with 23 platforms and 14 UAS framework components, answers a question the commercial market had been unable to answer for itself: which drones can government agencies, federal contractors, and critical infrastructure operators trust?[1] The answer carries weight that no amount of private certification could replicate, because it comes from the institution with the most comprehensive evaluation capability and the most direct stake in getting the answer right. The result is that Blue UAS approval has become, in practice, the national trust standard for drone procurement, reshaping investment patterns, forcing product redesigns, and creating a competitive moat for companies on the cleared list that their unapproved competitors cannot easily overcome.

The third channel is market direction. Defense procurement is, at industrial scale, the most powerful form of demand signal available in the American economy. A single DoD program of record, a formal commitment to procure a specified capability over a multi-year period, can justify the capital investment required to build a factory, train a workforce, and establish a supply chain. The companies that receive those commitments do not merely fulfill them. They become the anchor tenants of an industrial ecosystem: their facilities attract suppliers, their workforce needs pull from local training pipelines, and their procurement requirements define the standards that every company in their supply chain must meet.

Dual-Use Technology: The Innovation Loop Runs Both Ways

For most of the history of American aviation, the flow of innovation ran in one direction. Military investment in radar, jet engines, composite materials, GPS navigation, and miniaturized sensors created capabilities that eventually migrated into commercial products, sometimes decades after their defense origins. The technology transfer was real and consequential, but it was also unidirectional and slow, governed by classification requirements and procurement timelines that the commercial market had no mechanism to accelerate.

That model has been disrupted by the speed of commercial technology development in the autonomy era. Artificial intelligence (AI)-powered navigation, computer vision, obstacle avoidance, and swarm coordination are advancing faster in the commercial sector, driven by the competitive pressures of venture-backed startups, the open research ecosystem of academic AI, and the massive scale of commercial data collection, than traditional defense acquisition cycles can absorb. The DoD, for the first time in its history, finds itself in the position of a technology follower in some of the most strategically important domains of military aviation. Its response has been institutionally creative: rather than trying to accelerate its own internal development timelines, it has built the procurement mechanisms that allow it to purchase commercial innovation and adapt it to military requirements.

The dual-use dynamic is no longer a pipeline from defense to commercial. It is a loop, and the commercial sector is now the faster half.

This is precisely why the DIU's Other Transaction Authority (OTA) mechanisms and its pitch day model exist. OTA is a statutory authority that allows the DoD to contract for research, development, and prototype projects outside the Federal Acquisition Regulation, the dense, process-heavy rulebook that governs traditional defense contracting and that effectively excludes most commercial startups from competing for government business.[9] An OTA agreement can be negotiated in months rather than years. It can be structured to allow iterative prototyping rather than requiring a fully specified deliverable before work begins. The 369 proposals received in the DIU's 2025 Refresh Challenge are a measure of how effectively that openness has been communicated: companies from 19 nations submitted proposals not because they had navigated the acquisition bureaucracy but because the acquisition bureaucracy had, for once, come to them.

The result is a genuine technology exchange that now runs in both directions simultaneously. From defense to commercial: hardened communications infrastructure, endurance improvements, sensor miniaturization, and the operational discipline embedded in military-trained UAS professionals who transition to civilian careers. From commercial to defense: AI navigation stacks, computer vision platforms, rapid manufacturing methodologies, and software-first architectures that reduce the cost and timeline of fielding new capabilities. The dual-use dynamic is no longer a pipeline. It is a loop, and the policy frameworks, procurement mechanisms, and investment patterns of the drone industry are being reshaped by that reality.

Blue UAS: The Trust Framework That Reorganized the Market

The Blue UAS Cleared List was not, when the DIU first published it, conceived as a market-shaping instrument. It was conceived as a procurement tool, a practical answer to a practical problem. Federal agencies that wanted to replace their Chinese-manufactured drone fleets following the National Defense Authorization Act (NDAA) restrictions had no reliable way to identify which domestic alternatives met the security and performance standards that government operations require. Blue UAS provided that answer, and in doing so it produced consequences that extended far beyond the federal procurement problem it was designed to solve.

The August 2025 update to the Cleared List included 23 drone platforms and 14 UAS framework components, reflecting the growth of the domestic manufacturing base that NDAA procurement restrictions had stimulated.[1] The manufacturers on the list, Skydio, AeroVironment, Parrot, Teal, Teledyne FLIR, Shield AI, Inspired Flight, Freefly Systems, ModalAI, and others, represent the core of what has become a new competitive tier in the domestic drone market: companies whose Blue UAS approval functions as a trust credential that their unapproved competitors cannot replicate.

The commercial implications of that trust credential are significant and still unfolding. State and local government agencies that receive federal funding are increasingly required, under the Office of Management and Budget's November 2025 implementation memorandum, to comply with the same UAS procurement standards that apply to federal agencies directly.[4] Public safety departments that built their drone programs around DJI equipment are discovering that their access to federal grant funding depends on transitioning to Blue UAS-approved alternatives. Critical infrastructure operators, utilities, pipeline companies, airport authorities, are evaluating their own procurement standards against the Blue UAS framework, even where no federal requirement compels them to do so, because the reputational and liability implications of operating on non-cleared platforms have become difficult to ignore.

Blue UAS has become, in practice, the most consequential policy lever in the domestic drone market. It is not simply a list. It is a market-shaping instrument that simultaneously creates demand for approved manufacturers, imposes product redesign requirements that force supply chain sovereignty, and establishes a national trust baseline that commercial buyers, not merely government agencies, are beginning to use as a reference standard.

Replicator: A Manufacturing Demand Signal Unlike Any Before It

In August 2023, Deputy Secretary of Defense Kathleen Hicks announced the Replicator initiative, a plan to rapidly field thousands of attritable autonomous systems across multiple domains within 18 to 24 months.[2] The ambition was explicit: match and deter the autonomous systems capabilities that adversaries, particularly China, had been building for years. Small, inexpensive, proliferable autonomous systems can hold adversary forces at risk across contested environments in ways that larger, more expensive platforms cannot. If the United States military could field them at scale, on a timeline measured in months rather than years, it would represent a meaningful shift in the character of potential conflicts in the Pacific and beyond.

Replicator's production difficulties, well documented in subsequent defense analysis and congressional testimony, did not stem from a failure of strategic vision or a lack of funding. They stemmed from a structural gap that no amount of procurement urgency could bridge quickly: the United States did not have the domestic manufacturing capacity to produce attritable autonomous systems at the scale and cost the initiative required, on the timeline it demanded. The supply chain for small drone propulsion, battery cells, sensors, and electronics ran, in large part, through China, the same adversary the initiative was designed to deter.

Replicator did not simply create demand for drones. It revealed the gap between the demand the nation requires and the industrial capacity that currently exists to answer it.

The lesson of Replicator was absorbed directly into the design of the One Big Beautiful Bill Act's (OBBBA) $1.4 billion appropriation for the small UAS industrial base, a direct investment in production capacity rather than a procurement commitment that would have required manufacturers to self-fund the infrastructure needed to fulfill it.[8] By funding the capacity directly, the OBBBA attempted to solve the problem that Replicator could not solve through procurement urgency alone.

Skydio's SkyForge initiative, the $3.5 billion, five-year manufacturing expansion announced in April 2026, is the most visible private-sector response to the industrial capacity gap that Replicator made impossible to ignore.[3] SkyForge is not merely a factory expansion. It is an attempt to reconstruct the American drone supply chain from the component level upward: more than $1 billion directed specifically to domestic component suppliers, an invitation for those suppliers to co-locate with Skydio's expanded operations, and a production commitment backed by the $110 million Series F that institutional investors provided once the federal demand signal was clear enough to justify the capital deployment. Replicator revealed the gap. The OBBBA and SkyForge represent the first serious attempts to close it.

How Commercial Innovation Reaches the Pentagon

Traditional defense acquisition is not merely slow. It is structurally optimized for large, established defense primes with dedicated government contracting offices, cleared personnel, and the organizational bandwidth to manage the administrative requirements of Federal Acquisition Regulation-governed contracts over multi-year program timelines. For the drone and autonomy startup ecosystem, small, software-driven, commercially oriented companies whose product development cycles are measured in weeks, the traditional acquisition pathway is not a high bar to clear. It is a wall that most companies cannot reach the base of.

OTA mechanisms change this equation in ways that matter significantly for the commercial drone sector. An OTA agreement allows the DoD to contract for research, development, and prototype projects outside the Federal Acquisition Regulation framework. Contracting timelines that take years under traditional acquisition can be compressed to months under OTA. Prototyping agreements can be structured to allow iterative development, testing, feedback, revision, rather than requiring a fully specified deliverable at contract signature. Startups without government contracting infrastructure can compete for OTA opportunities on terms calibrated to their organizational reality rather than the requirements of a defense prime.

The DIU's Refresh Challenges and pitch day model are the most visible expression of the OTA pathway in the drone sector. By publicly soliciting commercial technology demonstrations, evaluating them against operational requirements, and channeling the most promising into OTA prototype agreements and eventually programs of record, the DIU has created a funnel through which commercial innovation enters the defense acquisition system at a pace that traditional contracting could not accommodate.

For drone companies contemplating their relationship with the DoD, the OTA pathway is not merely a financing option. It is a strategic posture. A company that has demonstrated its technology in an OTA prototype agreement and earned performance data in a military operational environment has established the kind of credibility that no commercial reference customer can replicate. The program of record that may follow an OTA prototype is, for a company with the manufacturing capacity to fulfill it, the demand signal that justifies the capital investment in domestic production infrastructure that Blue UAS approval and NDAA compliance require.

The Military Workforce: A Shared National Asset

The military UAS workforce is one of the most consequential and least discussed influences on the commercial drone sector. Every year, tens of thousands of service members separate from the armed forces carrying operational experience, technical training, and professional discipline that the commercial drone market urgently needs and cannot replicate through conventional hiring and training processes. The gap between the supply of this talent and the industry's ability to capture it represents one of the most immediately addressable workforce inefficiencies in the entire UAS ecosystem.

Army Military Occupational Specialty (MOS) 15W, Unmanned Aerial Vehicle (UAV) Operator, is among the most directly translatable military career paths to commercial UAS work. A 15W operator separating from active duty carries demonstrated competency in operating complex uncrewed systems under dynamic conditions; managing payload sensors and interpreting their outputs; navigating airspace regulations of significant complexity; maintaining detailed flight logs and equipment maintenance records; and bearing full operational responsibility for missions where failure has real consequences.[6] These are not adjacent skills to the core competencies of a commercial UAS professional. They are those competencies, developed at government expense, validated in operational environments, and ready to deploy the day after separation paperwork clears. Army MOS 15E (UAS repairer) and MOS 153D (UAS pilot warrant officer) extend the pipeline further, as do the Navy and Air Force equivalents whose skills map with equal directness onto the commercial sector's most acute talent shortages.

The barriers to capturing this talent are not technical. The Part 107 credential process is straightforward for veterans with active UAS experience: the airspace knowledge, systems familiarity, and operational discipline the exam tests are competencies military UAS professionals apply routinely. The salary ceiling on the commercial side, $100,000 to $180,000 or more for experienced operators in senior commercial or supervisory roles, is competitive with civilian career alternatives that typically require years of additional education to reach. The barriers are structural: the commercial drone industry and the military separation system are not communicating effectively enough to connect available talent with available opportunity before service members make their civilian career decisions.[6]

When military UAS operators, maintenance technicians, and program managers transition to civilian roles, they become the organizational substrate of the commercial industry: the program managers who know how to run multi-aircraft operations across wide geographic areas; the safety officers who understand what redundancy and reliability mean in systems whose failure has operational consequences; the startup founders who know from operational experience what problems are worth solving; and the instructors who can translate the gap between Part 107 certification and BVLOS operational readiness into a curriculum that produces mission-ready professionals. The two workforces, military and commercial, are not parallel tracks. They are interdependent components of the same national aviation capability.

The Strategic Takeaway: Defense Shapes the Entire Ecosystem

Across procurement, policy, manufacturing, and workforce development, the Department of Defense is the single most influential actor in the U.S. uncrewed aviation landscape. Its influence is not always visible in the commercial sector's press release record, but it is present in virtually every significant development of the past five years: in the NDAA restrictions that created the demand signal for domestic manufacturing; in the Blue UAS framework that reorganized the trust architecture of the entire drone market; in the Replicator initiative that revealed the manufacturing capacity gap that the OBBBA was designed to address; in the OTA mechanisms that have opened the Pentagon to commercial innovation at a pace traditional acquisition could never achieve; and in the workforce pipeline that produces trained UAS professionals at government expense and releases them into the commercial market every year.

The dual-use dynamic that defines this relationship is accelerating. As autonomy software, AI-powered sensor fusion, and machine learning-driven navigation become more central to both military and commercial drone operations, the distinction between a defense-grade capability and a commercial-grade capability is narrowing. The same detect-and-avoid system that enables a BVLOS infrastructure inspection over a power transmission corridor is a version of the same technology that enables a military intelligence, surveillance, and reconnaissance (ISR) platform to navigate contested airspace. The same supply chain sovereignty that the DoD requires for NDAA-compliant procurement is the supply chain sovereignty that the commercial industry needs to avoid the economic coercion that Chinese authorities demonstrated when they targeted Skydio's battery supply in October 2024.

The companies that will define the drone industry's next decade are the ones building with this dual-use reality as a founding assumption rather than an afterthought. They are structuring their product architectures for allied interoperability from the first line of code. They are pursuing Blue UAS approval not as a compliance requirement but as a commercial positioning strategy. They are building domestic supply chains not because the NDAA requires it but because they understand that a supply chain running through an adversary nation is not a supply chain at all. They are developing relationships with the DIU and with military program offices not as a parallel business development track but as the primary validation pathway for technology that will eventually serve both military and commercial markets. The future of uncrewed aviation will not be built by defense or commercial actors alone. It will be built in the space between them.

Looking Ahead — Chapter Twelve

Defense investment, for all its influence, operates within a larger geography: the regional economies that will either capture or miss the long-term economic value the drone industry creates. Chapter Twelve examines that geography, the states, cities, and corridors positioning themselves for drone economy leadership, and the investment strategies, workforce pipelines, and infrastructure commitments that separate the regions that will host the drone industry's next chapter from the ones that will watch it develop somewhere else.

Notes and Sources

1. Defense Innovation Unit (DIU). Blue UAS Cleared List, updated August 2025. Lists 23 cleared drone platforms and 14 UAS framework components; 369 proposals from 19 nations received in the 2025 Refresh Challenge. Primary source for Blue UAS framework discussion throughout this chapter.

2. U.S. Department of Defense. Replicator Initiative announcement, August 2023. Deputy Secretary of Defense Kathleen Hicks. Program objective: field thousands of attritable autonomous systems across multiple domains within 18–24 months. Source for manufacturing gap analysis.

3. Skydio, Inc. SkyForge Program Announcement, April 2026. $3.5 billion five-year manufacturing expansion; $110 million Series F at $4.4 billion valuation; 2,000+ direct manufacturing jobs; $1 billion+ directed to domestic suppliers. Source for manufacturing response discussion.

4. Office of Management and Budget. Government-Wide Implementation Memorandum: Secure UAS Procurement Requirements. November 21, 2025. Extends federal secure UAS procurement requirements to all federal agencies and recipients of federal funds, including state and local governments operating on federal grants.

5. Federal Communications Commission. National Security Covered List, updated December 2025. DJI added following FY2025 NDAA review; new products blocked from FCC radio frequency authorization. Source for market disruption context.

6. Military Transition Toolkit. Army MOS 15W UAV Operator to Commercial Drone Pilot Career Pathway Guide. January 19, 2026. Source for military-to-commercial workforce transition analysis, skills transfer assessment, and salary range data.

7. National Defense Authorization Act for Fiscal Year 2024 (Public Law 118-31). December 22, 2023. American Security Drone Act provisions; FAR interim rule effective November 12, 2024. Primary legislative authority for federal drone procurement restrictions creating Blue UAS demand signal.

8. One Big Beautiful Bill Act (OBBBA). Signed July 4, 2025. $1.4 billion appropriation for small UAS industrial base; designed to address manufacturing capacity gap identified through Replicator implementation.

9. Department of Defense / Defense Innovation Unit. Other Transaction Authority overview and DIU Pitch Day program documentation. Source for OTA mechanism description and commercial innovation procurement pathway analysis.

10. U.S. Air Force. Collaborative Combat Aircraft (CCA) program selection of Shield AI Hivemind autonomy platform. Source for technology validation channel discussion and dual-use dynamic analysis.

Economic Development and the Regional Drone Economy

"The drone economy is not being built on a national blueprint. It is being built in county commission meetings and airport authority boardrooms and community college curriculum committees, by the local leaders who understood, early, that the industry of the next decade was choosing its geography right now."

— *Sharon Rossmark*

The Pitch That Signals a Shift

She has 22 minutes on the agenda, and she uses every second. The economic development director from a mid-sized Southeast city is pitching a domestic drone manufacturer's site selection team, and she has not led with tax incentives. She has led with workforce: the community college drone technology program that graduated its first 34 students in May, the FAA-designated test site 38 miles north, the former automotive supplier with 120,000 square feet of precision manufacturing space sitting largely idle, and the cost of living that means a drone technician earning $58,000 a year can buy a house. By the time she finishes, the site selection team has asked for a second meeting. She already knows what the answer will be.

The regional drone economy is competitive, strategic, and increasingly defined by the communities that understand the moment and the ones that do not. Economic development has always been a competition for the right to host the industries of the future. Automotive manufacturing. Semiconductor fabrication. Digital technology. Each reshaped the regions that secured them and left behind the ones that waited too long or invested too little. Uncrewed aviation is the next iteration of that race, and it is being run right now in state capitals, county commission meetings, and regional airport authority boardrooms across the country.

The stakes are high and the window is not indefinitely open. The companies building the American drone industry are making location decisions now, decisions about where to site manufacturing operations, where to establish testing infrastructure, where to recruit from, and where to build the long-term supplier relationships that anchor industrial clusters for decades. The communities that have prepared for this moment are winning those decisions. The ones that have not are watching the announcements come from somewhere else.

This chapter is for the regions still in the race, and for the ones that think they have already lost it. The geography of the drone economy is not yet fixed. The industry is growing faster than any single region can absorb, which means that the competitive opportunity remains genuine for communities that are willing to act deliberately, invest strategically, and understand what drone manufacturers, operators, and researchers are actually looking for when they evaluate a location.

The States That Moved First

The competitive landscape of the regional drone economy did not emerge from a vacuum. It was built, deliberately, over years, by states that made early bets on uncrewed aviation as a strategic economic priority and then sustained those bets through budget cycles, political transitions, and years of development that produced no immediate return. The states that are now industry anchors are not lucky. They are early. And understanding what they did, specifically, sequentially, and at what cost, is the clearest available guide for the communities that are entering the competition now.

North Dakota's story has been told in detail in Chapter 9, but its regional development dimensions deserve emphasis here. The Northern Plains UAS Test Site, one of the FAA's seven federally designated UAS test sites, was not the end of North Dakota's investment. It was the anchor around which more than a decade of ecosystem development accumulated. University research programs at the University of North Dakota aligned their aviation curriculum with the test site's operational priorities. Workforce training pipelines were built before they were needed rather than after the gap became visible. The Vantis ground-based radar network, funded by more than $100 million in cumulative state investment, created BVLOS-capable airspace coverage across 56,000 square miles, the largest contiguous BVLOS airspace in the United States, and gave North Dakota something no other state could offer: a regulatory environment where the most commercially valuable class of drone operations could be routinely conducted, tested, and refined.[1] The $100 million investment leveraged

$400 million in federal and private capital.[2] That four-to-one return ratio is the arithmetic of patient, coordinated, compounding investment made before the market required it.

The states winning the drone economy competition did not wait for the industry to arrive. They built the conditions that made arrival inevitable.

Virginia invested heavily in drone delivery infrastructure, hosting Wing's commercial delivery operations and partnering with NASA's Langley Research Center on airspace management research.[6] The state positioned itself as a proving ground for urban and suburban drone logistics by creating the regulatory flexibility and federal research partnership that drone delivery companies needed to move from proof-of-concept to operational deployment. Virginia's proximity to the federal government, its abundance of defense contractors, federal agencies, and national security researchers, gave it a natural advantage in dual-use applications that the state exploited deliberately rather than accidentally.

Nevada and Texas leveraged favorable regulatory environments, abundant test airspace, and aggressive economic development incentives to attract manufacturers, operators, and autonomy companies seeking the room to test and scale that congested coastal airspace cannot provide. Nevada's designation as the site of the American Robotics Institute's operations and Texas's cultivation of a drone corridor along the Gulf Coast energy infrastructure represent different versions of the same strategic logic: identify the specific operational need that the drone industry's most valuable segments require, and build the infrastructure that meets it before the competition does.

Ohio's competitive advantage is rooted in a manufacturing base that the drone industry needs and other states cannot replicate quickly. The advanced manufacturing infrastructure built over decades of automotive supply chain development, precision machining capacity, quality control systems, logistics expertise, and a technically trained workforce, translates directly to drone component production. Communities that once depended on automotive jobs are discovering that the precision tolerances, material specifications, and assembly disciplines required by drone manufacturing map closely onto the capabilities already embedded in their workforce and their facilities. Ohio did not build a drone economy from scratch. It recognized that a drone economy was being built, and that the competitive advantage its manufacturing base represented was temporary if it was not activated deliberately.

The FAA Test Site Network: Regional Anchors with National Impact

The FAA's seven federally designated UAS test sites are among the most consequential pieces of public investment infrastructure in the drone economy, and among the least understood by communities that are not already in their orbit. They are not merely testing facilities. They are economic anchors: nodes around which workforce development, research partnerships, industry recruitment, and regulatory innovation cluster in ways that produce compounding returns over time.

The test sites provide something that no amount of private investment can replicate: a formal relationship with the FAA that allows experimental operations, BVLOS missions, autonomous systems testing, urban air mobility research, to be conducted in ways that would require lengthy waiver processes anywhere else.4 Companies choose to locate near test sites not primarily for the infrastructure they offer but for the regulatory relationship they embody. Proximity to a test site means proximity to the federal aviation authority's evaluation process, faster feedback loops, more direct access to the regulatory intelligence that determines what a company's next product needs to demonstrate, and the credibility that comes from testing in a federally recognized environment rather than a self-certified one.

The seven sites, in Alaska, Nevada, New York, North Dakota, Texas, Virginia, and Kansas, represent a deliberate geographic distribution designed to cover different operational environments, climate conditions, and use-case contexts. Alaska's Pan Pacific UAS Test Range offers something no other site can: extreme weather testing, vast uninhabited terrain, and the remote logistics delivery conditions that make it the most realistic environment in the country for long-range autonomous operations. Texas, anchored by the Lone Star UAS Center of Excellence and Innovation in Corpus Christi, leverages Gulf Coast energy infrastructure, expansive uncongested airspace, and year-round flying conditions that give it a natural advantage in infrastructure inspection, maritime surveillance, and energy sector applications. That distribution creates opportunity for regions not in proximity to a current test site: the FAA's framework anticipates expansion, and the communities that make the strongest case for their operational uniqueness, their workforce infrastructure, and their industry partnerships will be best positioned to make that case when the opportunity arises.[5]

Drone Corridors: The Infrastructure Investment That Defines the Next Decade

If test sites are the anchor points of the regional drone economy, drone corridors are the connective tissue that will determine how economic value flows between them. Designated BVLOS airspace routes, the aerial equivalent of interstates, are emerging as the next major category of aviation infrastructure investment, and the communities that position themselves along them early will capture the economic activity that follows, just as the communities along the Interstate Highway System captured the logistics and manufacturing activity that defined mid-century American economic geography.

Drone corridors are the interstates of low-altitude airspace. The communities that position themselves along them now are making a bet on geography that will pay for decades.

The Interstate Highway System did not merely enable faster travel between existing destinations. It created new economic geography: distribution centers, industrial parks, retail clusters, and residential communities that would not have existed without the infrastructure that made them viable. Drone corridors will do the same thing at the scale of low-altitude airspace. A corridor running along a utility transmission right-of-way through a rural agricultural region does not merely enable infrastructure inspection. It creates the foundation for precision agriculture operations, emergency medical supply delivery, and rural logistics services whose business models depend on the predictable, legally authorized BVLOS access that a designated corridor provides.

Several regional corridor initiatives are already underway. The New York UAS Corridor, developed in partnership with state agencies and the FAA, is designed to enable routine commercial drone operations across specific geographic routes through the Hudson Valley. Texas has invested in corridor development along Gulf Coast energy infrastructure, enabling the pipeline and power line inspection operations that represent some of the strongest commercial use cases for BVLOS operations. North Dakota's Vantis network, discussed in Chapter 9, is effectively a statewide corridor infrastructure investment, the most advanced in the country, that has already attracted companies whose business models depend on the routine BVLOS access it enables.

The strategic imperative for communities that are not yet part of a designated corridor is to get ahead of the designation process rather than react to it. Corridor routes are being planned now, based on operational demand, airspace geometry, radar coverage, and existing infrastructure. The communities that understand this planning process, engage with the FAA's relevant program offices, and make the case for corridor routes that serve their operational geography will influence where the infrastructure lands. The ones that wait for the corridor map to be published will inherit the decisions that others made.

The Rural Opportunity: Where Drones Matter Most

There is a persistent misconception in the public conversation about the drone economy that frames it primarily as an urban story: delivery drones navigating city skies, air taxis connecting metropolitan airports, autonomous surveillance systems monitoring dense urban infrastructure. The reality, in terms of economic impact, near-term commercial viability, and the density of unmet need that drone technology addresses, is nearly the inverse. The most transformative applications of drone technology are disproportionately concentrated in rural communities, and the economic development opportunity those applications represent is, for many rural regions, the most significant industrial opportunity they have encountered in a generation.

Precision agriculture is the clearest expression of this reality. The ability to map crop health, soil moisture, and pest pressure across thousands of acres in a single flight cycle, generating granular, field-level data that allows farmers to prescribe inputs at the plant level rather than the field level, is not incrementally valuable in agricultural communities. It is transformative. The operational savings from precision application of fertilizers and pesticides, the yield improvements from early identification of crop stress, and the labor cost reductions from drone-based monitoring and spraying are, in aggregate, large enough to materially affect the economics of farming operations that are otherwise being squeezed by commodity price pressure and rising input costs. And the agricultural drone service providers, maintenance technicians, and data analysts who make these operations possible are jobs that do not require a metropolitan labor market. They require proximity to the farms being served.

Rural infrastructure inspection is equally compelling as an economic development driver. The United States has more than 600,000 bridges, hundreds of thousands of miles of transmission line and gas pipeline, and a rural road network that requires continuous monitoring at a cost and frequency that ground-based methods cannot sustain. Drone-based inspection is faster, cheaper, and safer than the alternatives, and the inspection work follows the infrastructure, which means that the jobs, the companies, and the economic activity follow it too. A rural county with a gas pipeline corridor running through it, or a transmission line right-of-way, or a cluster of aging bridges carrying agricultural freight to market has the raw material of a drone services economy embedded in its geography. The question is whether it has the workforce and the business infrastructure to capture the value.

Rural drone manufacturing represents a different but equally real opportunity. Drone manufacturing and component production are labor-intensive, technically skilled, and geographically flexible in ways that differentiate them from the coastal-biased industries that have dominated the recent technology economy.[7] The precision assembly work required to build a commercial drone airframe, the electrical harness installation and quality testing that drone production requires, and the component machining and fabrication that feeds the assembly line are all skills that exist, in varying degrees, in rural manufacturing communities across the Midwest, the Southeast, and the rural West. A community that graduates UAS maintenance technicians from its local technical college, has idle precision manufacturing space available at below-market lease rates, and offers a cost of living that makes a manufacturing wage genuinely competitive with coastal alternatives is a credible candidate for drone manufacturing investment.

Workforce as the New Competitive Advantage

The economic development director in the chapter's opening vignette understood something that her counterparts in less successful regions had not yet internalized: workforce is not a supporting consideration in drone industry site selection. It is the primary criterion. The companies making location decisions in the commercial drone sector are not choosing between communities based on tax incentives alone. They are choosing between communities based on whether a trained, credentialed, locally available workforce exists that can staff their operations from day one, or whether training that workforce from scratch would require years of investment that their business timeline cannot absorb.

The community that can demonstrate a functioning UAS workforce pipeline, a community college running a FAA UAS Collegiate Training Initiative (CTI)-recognized program, a technical school producing maintenance technicians credentialed to ASTM International F3600 standards, an industry advisory board connecting curriculum to operational requirements, and a track record of graduates placed in local employment, has a competitive advantage that a larger tax incentive package from a community without those assets cannot overcome.9 Drone companies are not looking for the lowest cost location. They are looking for the lowest risk location, and the risk most difficult to manage is the workforce risk.

The cost-of-living dimension of this calculation deserves explicit attention. The drone industry's workforce economics are more favorable to mid-sized and rural communities than those of almost any comparable technology sector. A drone operations manager earning $75,000 in a mid-sized Southeast city is in a materially stronger financial position than the same professional earning $120,000 in San Francisco or Seattle. This is not a niche consideration. It is one of the most frequently cited factors in the career decision-making of the mid-career professionals, average age 39 for women entering the certified commercial drone pilot workforce in 2025, who are choosing where to build their drone industry careers.[8]

The Economic Multiplier: Why Drone Clusters Compound

The economic development case for pursuing drone industry clusters is not merely about the direct employment a drone company brings to a community. It is about the multiplier effects, the supplier development, the service industry expansion, the tax base growth, and the innovation ecosystem formation, that a drone cluster generates when it reaches the scale and duration required to anchor a regional economy. These effects have been documented across every major industrial cluster in American economic history, from the automotive ecosystem of the Midwest to the semiconductor cluster of Silicon Valley to the biotech corridor of the Boston Route 128 corridor. The drone industry, as it matures into the infrastructure and defense markets that will define its scale, will generate comparable effects in the regions that host its most significant operations.

The primary effects are the most visible: direct employment in manufacturing, operations, engineering, data analysis, maintenance, and management at the drone companies and service providers that constitute the cluster's core. Capital investment in facilities, equipment, and infrastructure that flows from those companies into the regional construction, real estate, and equipment supply markets. Federal research and development funding that follows companies with strong federal relationships into the universities and research institutions they partner with locally.

The secondary effects are where the compounding begins. A drone manufacturing facility that requires precision metal components creates demand for local machining shops. A drone services company that operates multiple aircraft requires local fuel, maintenance, and logistics services. An autonomy software developer that employs 50 engineers requires housing, childcare, restaurants, healthcare, and the full supporting infrastructure of a growing professional workforce. In economically distressed communities where the commercial and service sectors have contracted alongside the manufacturing base, the multiplier effect of a drone industry cluster can be disproportionately large relative to the direct employment count, precisely because the anchor employer's spending flows into a local economy whose service capacity is well below what the demand would support.

The tertiary effects, university partnerships, research program formation, innovation ecosystem development, are the ones that determine whether a drone cluster becomes a durable regional asset or a single-generation economic event. North Dakota's University of North Dakota aviation program, Virginia's NASA Langley partnership, Ohio's manufacturing research center connections, each represents a community's decision to treat its drone industry presence as a long-term ecosystem investment rather than a short-term economic development win.

A Framework for Regions Entering the Race

For communities that have not yet begun building their drone economy strategy, the window is still open, but it is narrowing, and the communities entering the competition now face a more demanding bar than those that entered five years ago. The early movers had the advantage of building before the competition was fully formed. Communities entering today are building against a competitive landscape that already has established players, and they will need to be specific, strategic, and credible in their differentiation to attract the attention of companies that now have multiple proven locations to evaluate.

The first step is asset inventory, not aspiration. Communities that begin their drone economy strategy with a list of economic development wishes rather than an honest accounting of existing assets routinely waste the early months of their strategy on positioning that the market cannot validate. The honest question is not "what do we wish we had?" but "what do we already have that the drone industry needs?" A regional airport with a reasonably uncongested flight path environment. A community college with an aviation or electronics program that could be extended into UAS with targeted investment. A former manufacturing facility with the floor space, the electrical infrastructure, and the loading dock access that drone production requires. A workforce with precision manufacturing experience from an adjacent industry. These existing assets, assessed candidly, mapped to specific drone industry needs, and packaged in a way that site selection teams can evaluate, are the foundation of a credible strategy.

The second step is pipeline investment, not pipeline announcement. The communities that are winning drone industry projects do not merely have UAS training programs. They have UAS training programs with graduates, with employer relationships, with placement data, and with the demonstrated operational readiness that distinguishes a functioning pipeline from a planned one.[11] The community college that has been running a Part 107 preparation course for 18 months and placed 60 percent of its graduates in local drone operator roles has something that a community college that has just submitted a FAA CTI application does not: evidence. That evidence is what converts a site selection team's interest into a commitment.

The third step is federal engagement, not just federal funding pursuit. The communities that have captured the most drone industry economic value are the ones that understood the federal program architecture, the FAA test site network, the EDA regional innovation grants, the WIOA workforce development framework, the DoD procurement pipeline, and positioned themselves within it deliberately rather than pursuing individual grant opportunities in isolation.[10] Communities that invest in the organizational capacity to navigate that complexity, a grants manager, a federal relations coordinator, a regional coalition with the bandwidth to pursue multiple programs simultaneously, are converting the federal fragmentation problem into a competitive advantage over communities that cannot absorb the administrative burden of multi-agency engagement.

The fourth step is marketing specificity, not marketing volume. The drone industry's site selection process is driven by technical and operational criteria, not by general economic development marketing. A community that can tell a specific, evidence-based story about its operational environment, the airspace characteristics that make BVLOS testing feasible, the agricultural sector that makes precision agriculture drone services economically viable, the infrastructure corridor that generates year-round inspection demand, will be heard by the engineers and operations managers who actually drive location decisions in ways that a community promoting its general business-friendliness will not.

The Strategic Takeaway: The Drone Economy Is Local

Uncrewed aviation may be a national industry story, shaped by federal legislation, defense procurement, international competition, and the capital markets of coastal financial centers, but its economic impact is fundamentally regional. The jobs, the facilities, the tax revenue, the workforce development, and the supplier relationships that define the drone industry's long-term economic contribution to American prosperity will land in specific communities, on specific parcels of land, near specific airports and test sites and community colleges. They will land where the conditions for drone industry success have been deliberately built, not where they have been passively hoped for.

The states that moved early, North Dakota, Virginia, Nevada, Texas, Ohio, and the others that made early, patient, compounding investments in test site infrastructure, workforce pipelines, regulatory relationships, and industry recruitment, have established positions that will be difficult to displace. They did not become drone economy leaders because they were fortunate. They became drone economy leaders because their decision-makers understood, earlier than their competitors, that the industry of the next decade was choosing its geography right now, and that the window for positioning was open, but not permanently so.

The drone economy will not be evenly distributed. It will concentrate in the regions that understood the moment and acted decisively. The race is underway.

The competitive opportunity remains genuine for communities entering the race. The industry is large enough, and growing fast enough, to require a geographic footprint that no existing cluster of states can fully absorb. As the BVLOS regulatory framework moves toward implementation, it will create demand for operational infrastructure in markets that existing clusters cannot serve from their current locations. The manufacturing expansion represented by Skydio's SkyForge commitment and the broader OBBBA industrial base investment will create supplier development opportunities in communities adjacent to those expansions. The military workforce transition pipeline will deposit trained UAS professionals in communities near military installations across the country, giving those communities a talent base to build around that did not require them to build the training pipeline from scratch.

The drone economy is not being built on a national blueprint. It is being built in county commission meetings and airport authority boardrooms and community college curriculum committees, by the local leaders who understood, early, that the industry of the next decade was choosing its geography right now. The regions that want to be on the right side of that choice have the tools, the frameworks, and the federal investment architecture to make it happen. What they need is the strategic clarity to understand what they are actually competing for, the organizational capacity to pursue it systematically, and the patience to sustain the investment through the years between commitment and return. The race is underway. Some regions are already sprinting. Others are just now lacing their shoes. The finish line is far enough away that neither group has yet determined the final standings, but the gap between them is widening every quarter.

Looking Ahead — Part IV

Part IV turns from diagnosis to synthesis: the policy recommendations, industry strategies, and institutional investments that the book's full argument points toward. Chapter Thirteen examines Advanced Air Mobility, the certification races underway, and the infrastructure and workforce questions that will determine whether eVTOL transforms urban transportation or remains a premium novelty.

Notes and Sources

1. North Dakota Legislature. House Bill 1038, signed February 2025. $11 million appropriation for integration of Northern Plains UAS Test Site Vantis radar platform with FAA Pathfinder program, expanding BVLOS-capable coverage from 3,000 to 56,000 square miles. Source for North Dakota competitive advantage analysis.

2. Michael Standaert. "Working with FAA expands North Dakota commercial drone prospects." North Dakota Monitor, February 28, 2025. Reports on Vantis-Pathfinder integration, $100 million cumulative state investment, and $400 million in leveraged federal and private capital.

3. Emma Larson. "North Dakota Just Became the Drone Capital with FAA Data." KVRR / North Dakota Real Estate & Business, April 19, 2026. Governor Burgum characterization of FAA data agreement; R&D acceleration estimates; development cost savings for companies in the state.

4. Federal Aviation Administration. UAS Test Site Program overview and designated site locations. FAA.gov, accessed May 2026. Source for test site geographic distribution, operational authorization scope, and economic anchor function.

5. FAA Reauthorization Act of 2024, Public Law 118-63, Section 925. $6 million per year (FY2025–FY2028) for FAA-designated UAS test sites, structured as matching funds for commercial technology demonstration.

6. Wing Aviation LLC. Commercial drone delivery operations overview. Wing.com, accessed May 2026. Source for Virginia drone delivery infrastructure context.

7. Skydio, Inc. SkyForge Program Announcement, April 2026. $3.5 billion manufacturing expansion; domestic supplier co-location program; 5,000+ projected jobs. Source for manufacturing investment opportunity in rural and post-industrial communities.

8. Federal Aviation Administration. Civil Airmen Statistics, December 31, 2025. FAA Aeronautical Center, Oklahoma City, OK. Average age of women entering the commercial drone pilot workforce: approximately 39. Source for workforce cost-of-living analysis.

9. Unmanned Safety Institute (USI). UAS Maintenance Technician Program launch, February 2026. ASTM International F3600 standard; 12-week accelerated program at three community colleges. Source for workforce pipeline credentialing context.

10. U.S. Economic Development Administration. Regional innovation grant programs and drone corridor development investment framework. EDA.gov, accessed May 2026. Source for federal economic development funding context.

11. Federal Aviation Administration. UAS Collegiate Training Initiative (CTI) Five-Year Review, 2020–2025. April 2025. 145+ institutions; geographic distribution across all U.S. regions. Source for community college workforce pipeline discussion throughout.

PART IV: THE FUTURE

Future Flight

How Talent, Capital and Manufacturing will Shape the Next Era of Uncrewed Aviation

About Part IV

Part IV is where synthesis meets forward vision. The first three parts have established the technology landscape, diagnosed the workforce deficit, examined the manufacturing imperative, and mapped the investment ecosystem. Part IV builds on all of that to ask the harder questions: What is coming next in Advanced Air Mobility? Who will make the regulatory decisions that determine the airspace architecture of the next decade? What would a genuinely integrated national strategy look like? And ultimately, what must be built, by whom, with what urgency, to ensure that the United States leads the uncrewed aviation era rather than cedes it? Part IV is the book's argument at its most expansive, its most urgent, and its most hopeful.

Advanced Air Mobility and the Next Frontier

"Every era of aviation was built twice: once in the air and once in the institutions, regulations, and workforce pipelines that made sustained flight possible. Advanced Air Mobility is being built the same way, and the decade in which it gets built will determine who leads it for the next century."

— *Sharon Rossmark*

The Frontier Arrives Quietly

The flight lasts eight minutes. The same trip, by ground, takes 52 minutes on a good day and considerably longer in traffic. The passenger, who booked through an app the night before, boards at a vertiport on a downtown rooftop, sits in a four-seat cabin quieter than a helicopter, and watches the city fall away below. What she experiences is convenience, the ordinary magic of a technology that works. What she does not experience is the decade that preceded those eight minutes: the certification engineers who negotiated every performance standard with the FAA over years of technical review; the battery designers who pushed energy density into ranges that made commercial operations economically viable; the airspace managers who built the traffic management architecture that lets this aircraft share the urban sky with helicopters, commercial airlines, and thousands of small commercial drones; the workforce trainers who produced the pilots and technicians the operation required; and the regulatory negotiators who argued the legal framework into existence in the space between two aviation eras.

Advanced Air Mobility (AAM) is the next frontier of uncrewed aviation, and unlike the small UAS operations described in the first twelve chapters of this book, AAM is not simply being flown. It is being built, certified, and argued into existence in real time. The electric vertical takeoff and landing (eVTOL) aircraft that AAM companies are developing represent a genuine discontinuity in aviation history: the first new category of passenger-carrying aircraft to seek FAA type certification in decades, powered by electric propulsion architectures that have no precedent in the regulatory frameworks that govern traditional aviation. The decisions being made in the next three years, in certification offices, at vertiport construction sites, in airspace management standards bodies, and in the workforce training programs that will produce the pilots and technicians AAM requires, will determine whether the United States leads this sector or cedes it to the competitors who are building just as urgently.

Part IV begins here: with the future arriving quietly, eight minutes at a time, in the cities and corridors that will define the geography of the next aviation era.

What Advanced Air Mobility Actually Is, and What It Is Not

The public conversation about drones and Advanced Air Mobility often blurs distinctions that are, in regulatory and operational terms, foundational. The small UAS operations described across the first twelve chapters of this book, infrastructure inspection, precision agriculture, drone delivery, public safety, are primarily conducted by uncrewed, remotely operated or autonomous aircraft flying below 400 feet above ground level. They are regulated under Part 107 of the Federal Aviation Regulations, or under waivers and authorizations issued against that framework. They are categorized by the FAA as small unmanned aircraft systems (UAS): under 55 pounds, operating within a bounded regulatory envelope that has been progressively expanded since 2016.

AAM is different in kind, not merely in degree. The aircraft being developed by Joby Aviation, Archer Aviation, Wisk Aero, and their competitors are passenger-carrying vehicles, complex electromechanical systems designed to carry human beings through the urban air with the same safety standard that applies to commercial airliners. They are not regulated under Part 107. They are being certified under the FAA's Part 21 type certification process, the same rigorous, multi-year framework that governs every crewed aircraft that has ever received FAA approval.[11] They will operate under Part 135 air carrier certificates, not remote pilot certificates. Their pilots will hold conventional airman certificates, not commercial drone operator credentials. Their maintenance technicians will require qualifications analogous to the Airframe and Powerplant (A&P) certificates that govern traditional aircraft mechanics.

The distinction matters for this book's central argument because the workforce, the regulatory framework, the infrastructure, and the capital requirements for AAM differ substantially from those for small UAS, while the two sectors share the same airspace, many of the same technology suppliers, and the same national interest in American competitiveness that has animated every preceding chapter. AAM is not an extension of the drone economy. It is a new aviation sector built on electric propulsion, autonomy, and urban mobility, and it demands a workforce and infrastructure framework that does not yet fully exist.

The Certification Race: Who Is Closest to the Finish Line

FAA type certification for eVTOL aircraft follows the same five-stage process used for traditional crewed aircraft. Stage 1 establishes the conceptual design and initiates FAA engagement. Stage 2 defines the airworthiness requirements the aircraft must meet. Stage 3 develops the compliance plans that will demonstrate those requirements are met. Stage 4 executes the testing and analysis program that generates the compliance data. Stage 5 closes out the documentation and issues the final approval. The process is rigorous, multi-year, and unforgiving of capital miscalculation, as Lilium's insolvency in 2024 demonstrated with painful clarity. A company can build technically impressive hardware and still fail the certification race if its capital runway expires before its compliance data is complete.

Certification is the crucible. The companies that survive it will not merely have a product. They will have a franchise, the regulatory authority to carry passengers that no competitor can acquire quickly.

Joby Aviation is the current leader in the certification race by most publicly available measures. In late March 2026, the FAA confirmed that Joby had successfully completed Stage 4 of the five-stage type certification process, a critical airworthiness conformity review validating that the physical hardware matches the approved design, with propulsion system reliability and fly-by-wire redundancy among the key elements confirmed.

Stage 5 compliance flight testing, in which FAA pilots fly the production-representative aircraft and generate the independent compliance data required to finalize the certificate, is underway.[1] Joby is targeting a late 2026 commercial launch in partnership with Delta Air Lines, with initial service planned for New York City and Los Angeles, a timeline that reflects both the company's technical progress and the inherent uncertainty of a regulatory process whose final stages have never been executed for this category of aircraft. The S4's five-rotor tilting architecture, its 150-mile range, and its 200 mph cruise speed represent performance parameters that no traditional aircraft of comparable size could match.

Archer Aviation has achieved 100 percent Method of Compliance acceptance for its Midnight aircraft, a milestone meaning the FAA has agreed on the specific tests, analyses, and demonstrations that Archer must complete to show compliance with each applicable airworthiness requirement.[2] Archer is advancing through its certification stages with a target for type certification in the near term. Archer's strategic pivots in the years following the SPAC-era correction, the Anduril defense partnership, the United Airlines and Stellantis strategic investments, the acquisition of Lilium's patent portfolio, and the selection as the official air taxi provider for the 2028 Los Angeles Olympics, have created the multi-path revenue strategy and institutional stakeholder commitment that the SPAC era's pure commercial bet could not sustain.[4,5,12]

Wisk Aero is pursuing a fully autonomous certification pathway, an approach further from completion than the piloted strategies pursued by Joby and Archer, but that, if successful, would produce an aircraft that requires no pilot and therefore eliminates the most significant per-flight operating cost in the AAM business model.[10] The autonomous pathway requires the FAA to certify not only the aircraft's airworthiness but the reliability and fail-safety of its autonomous flight system, a standard the FAA has not previously applied to passenger-carrying vehicles. Wisk's approach is a longer-horizon bet, but the economics of a pilotless air taxi, if the certification can be achieved, are substantially more favorable than those of the piloted alternatives.

Beta Technologies (NYSE: BETA), headquartered in Burlington, Vermont, brings a fourth and strategically distinct approach to the certification race. Where Joby and Archer are pursuing full powered-lift type certificates for their eVTOL aircraft, Beta is executing a deliberate stepwise certification strategy: certifying its conventional takeoff and landing CX300 fixed-wing aircraft first, generating FAA compliance data and institutional familiarity in the process, then applying that foundation to accelerate certification of the ALIA 250 eVTOL. The CX300 projects type certification by late 2026 or early 2027; the ALIA 250 is targeted to follow approximately one year later. The stepwise logic is explicit: each certification effort informs the next, streamlining documentation, building continuity with FAA personnel, and reducing risk across programs. Beta surpassed 85,000 hours of flight and ground testing, made the United States' first passenger-carrying flight by an electric aircraft in June 2025, and was selected for 7 of 8 FAA eVTOL Integration Pilot Program sites, the most of any electric aircraft developer. Its customer base includes UPS and United Therapeutics for cargo and medical logistics applications; a $300 million strategic investment from GE Aerospace anchors its propulsion development. Beta's approach represents a different and credible theory of how to win the certification race: not by racing toward a single finish line, but by building the regulatory relationship incrementally, on a timeline that patient capital can sustain.[13]

Lilium's October 2024 insolvency is the cautionary data point that every remaining AAM company has internalized. Lilium built genuinely innovative technology, a novel ducted-fan architecture that attracted more than $1 billion in investment and secured high-profile orders. What it could not build fast enough was the compliance data that the certification process requires. The SPAC structure that provided its growth capital imposed quarterly reporting and liquidity expectations calibrated to public market investor timelines, timelines that are structurally incompatible with the years that aircraft certification requires. The company exhausted its capital before its compliance program was complete. The technology lived on in the 300 patents that Archer acquired for €18 million. The company did not.[3] AAM companies that survive the certification race will not do so on technology merits alone. They will do so because they structured their capital, their investor relationships, and their operational priorities around the non-negotiable timeline that certification imposes.

Vertiports and the Infrastructure Problem No One Can Ignore

The certification of the aircraft is the necessary condition for AAM operations. The infrastructure to support those operations is the sufficient condition, and it is, in many ways, the more difficult of the two challenges to solve. A vertiport is not simply a landing pad on a rooftop. It is a node in a complex physical and electrical system that must integrate power delivery, passenger processing, aircraft charging and turnaround, safety oversight, weather monitoring, and connectivity to both the airspace management system and the ground transportation network that passengers use to reach it. Building that infrastructure at the locations, in the quantities, and to the standards that commercially viable AAM operations require is a capital and logistics challenge whose scale the industry has not yet fully confronted.

The power infrastructure requirements alone represent a meaningful constraint on where and how quickly AAM can scale. Current estimates suggest that an early-stage vertiport serving a modest volume of eVTOL operations will require between two and five megawatts of electrical capacity, comparable to the demand of a small commercial district.[8] As operations scale, that requirement is projected to rise to between 10 and 50 megawatts per vertiport, depending on aircraft size, flight frequency, and charging architecture. The urban power grids that serve the downtown commercial districts and airport-adjacent sites that are most commercially attractive for vertiport placement were not designed for this incremental demand. Grid upgrades at the required locations will require coordination between vertiport developers, commercial real estate owners, utilities, and municipal authorities whose investment cycles and approval processes operate on timelines that may not align with the commercial launch schedules AAM companies are targeting.

The infrastructure race is as important as the certification race, and far less visible to the investors, regulators, and analysts tracking certification milestones. The aircraft that cannot land profitably anywhere has not won anything.

The builders entering the vertiport space represent a wide variety of institutional actors, and that variety is itself a signal of how genuinely infrastructure-dependent AAM is. Airports, particularly smaller regional airports seeking new revenue and relevance in an era of airline consolidation, are natural anchor tenants for vertiport development, with existing security infrastructure, air traffic control relationships, and regulatory familiarity that commercial real estate developers lack. Real estate developers are exploring vertiport integration into mixed-use developments, positioning rooftop aviation access as a premium amenity. Utility companies are examining the role of vertiport infrastructure in grid modernization strategies, recognizing that the high-density electrical load of a charging vertiport may serve as a useful anchor for distributed energy management systems they are already building. State and local governments are competing, in parallel to the regional drone economy competition described in Chapter 12, to attract the first vertiport developments as economic development wins with long-term transportation system implications.

Airspace Integration: The Systems-Level Challenge

AAM aircraft will not operate in isolation. They will operate in the same urban and suburban airspace simultaneously used by traditional crewed aviation, helicopter operations, commercial drone corridors, and the UAS Traffic Management (UTM) systems built to manage small UAS operations at scale. The airspace integration challenge is not merely a technical problem to be solved by sufficiently sophisticated software. It is a systems-level engineering, regulatory, and political challenge that requires coordinated action across stakeholders who have historically operated with minimal awareness of each other's requirements.

The fundamental question is how to define and manage the altitude band between approximately 400 feet and 5,000 feet above ground level, the space that AAM aircraft will primarily occupy and that intersects with both the low-altitude UAS operating environment and the approach and departure corridors of traditional airports.[6] The FAA's existing air traffic control architecture was not designed for the density of operations that AAM and commercial BVLOS drone operations will simultaneously generate in this altitude band. Voice-based control between human controllers and human pilots cannot scale to the volume of operations that AAM's commercial viability requires. Digital, automated traffic management is the only scalable answer, and the frameworks for that answer are being developed in parallel to the certification processes and infrastructure investments that AAM requires.[7]

The workforce implications of the airspace integration challenge deserve particular attention. Managing low-altitude traffic in the AAM era will require a new category of aviation professional whose skills sit at the intersection of traditional air traffic control, digital systems management, and the operational characteristics of both autonomous small UAS and piloted eVTOL aircraft. This profession does not yet have a defined credential pathway, a training curriculum, or a compensation structure. The FAA's Air Traffic Organization and the aviation workforce development community are beginning to grapple with this gap, but the timeline for producing certified low-altitude airspace managers in the numbers that commercial AAM operations will require is considerably shorter than the timeline that educational institution development and FAA credentialing processes typically require.

The Workforce AAM Requires, and Why It Doesn't Exist Yet

Every chapter of this book has returned, in some form, to the same structural constraint: the technology is advancing faster than the workforce that must deploy it. That constraint is present in small UAS, where the BVLOS regulatory window is opening faster than the training pipelines producing BVLOS-capable operators. It is present in drone manufacturing, where the domestic industrial capacity being built by SkyForge and the OBBBA investment requires technicians who do not yet exist in sufficient numbers. And it is present in AAM, where an entirely new category of aviation professional must be trained, certified, and deployed on a timeline that intersects with commercial launch schedules of companies that will need them ready before certification is complete.

The eVTOL pilot represents the most visible of the new workforce categories AAM requires, and the most complex to define. These professionals are not traditional airline pilots, the type ratings, training hours, and operational culture of Part 121 commercial aviation do not map cleanly onto the systems-management, high-automation cockpit environment of an eVTOL operation. They are not small UAS remote pilots, the certification requirements for passenger-carrying aircraft are categorically more demanding than those for commercial drone operations. They are a new hybrid role that requires the airspace awareness and regulatory literacy of a certificated pilot, the automation proficiency of an advanced systems operator, and the customer service orientation of a commercial transport professional. No existing training program produces this combination because no existing aircraft requires it. The training programs being developed by Joby, Archer, and their competitors are, by necessity, proprietary and aircraft-specific, which means that the certification of the aircraft and the credentialing of the pilots who fly it must advance in parallel, on a compressed timeline, with no established precedent to guide either process.

The eVTOL maintenance technician is arguably a more urgent near-term workforce gap than the pilot, precisely because maintenance technicians must be trained and credentialed before the aircraft enters commercial service while pilots can complete a portion of their training concurrently with the aircraft's own certification testing. eVTOL maintenance requires expertise in high-voltage electrical systems, the 400 to 800 volt battery architectures that power eVTOL propulsion represent energy levels that require specialized safety training and tooling that traditional aviation mechanics do not routinely encounter. It requires familiarity with composite airframe materials, additive manufacturing components, and the novel structural configurations that eVTOL designs employ. The FAA's A&P certificate provides the foundational credential, but the eVTOL-specific type training that supplements it must be developed and approved by the FAA for each aircraft type, a process that is underway but not yet complete for any commercial eVTOL platform.

The vertiport operator is a profession that does not yet exist at all. Ground operations at a busy vertiport will require the simultaneous management of aircraft charging and turnaround, passenger boarding and deboarding, safety oversight, weather monitoring and flight advisory communication, coordination with airspace management systems, and emergency response readiness. These functions draw on elements of airport ground operations, customer service management, electrical systems oversight, and air traffic coordination, but no existing professional certification covers their combination. The institutions that define this credential first, build the curriculum around it, and produce the first cohort of certified vertiport operators will have established themselves as indispensable partners in the AAM industry's commercial launch, a position analogous to the community colleges that built drone pilot training programs before the commercial market required them and found themselves with a structural advantage when the demand materialized.

Community Impact and the Politics of the Urban Sky

Advanced Air Mobility raises questions about urban airspace, community impact, and public acceptance that small UAS operations have largely avoided. A commercial drone delivering a package weighs a few pounds, operates at low altitude for a few minutes, and affects the immediate delivery area. A commercial eVTOL aircraft carrying four passengers operates at higher altitudes, on fixed corridor routes, at frequencies that will generate persistent noise in the neighborhoods below its flight paths and the communities adjacent to its vertiports. The public policy questions this raises are not technical. They are political, and their resolution will shape the pace and geography of AAM deployment as significantly as any engineering milestone.

The noise question is the most immediate. Contemporary eVTOL designs are quieter than helicopters, in some configurations significantly so, but quieter than a helicopter is not the same as quiet. The operational noise profile of a vertiport with 50 daily aircraft movements generates a community experience that neighbors will have opinions about, city councils will regulate, and planning commissions will weigh against the economic development arguments that vertiport developers will advance.[9] The noise certification standards the FAA is developing for eVTOL aircraft are an early attempt to define the technical parameters within which community noise must be managed, but the standards will not resolve the underlying political negotiation between commercial aviation interests and the residents of the neighborhoods where AAM operations will be most commercially attractive.

The economics of early AAM operations will almost certainly produce a pricing structure that makes the service accessible primarily to business travelers and premium commuters willing to pay a significant premium over ground transportation alternatives. The arc from premium service to accessible transportation is the trajectory that every successful transportation technology has followed. But the communities that will live under the flight paths and adjacent to the vertiports that serve early AAM operations have legitimate interests in the planning decisions that determine where AAM infrastructure is sited. The cities and metropolitan planning bodies that engage these communities seriously, early, and with genuine responsiveness to their concerns will build the political durability that AAM operations in dense urban environments will require.

The workforce development dimension of AAM's community impact story deserves explicit attention. The pilots, technicians, vertiport operators, and airspace managers that AAM requires are jobs that could be held by residents of the communities where AAM operates, if those residents have access to the training, credentialing, and hiring pipelines that connect them to the opportunity. The communities that are positioned to capture not just the transportation service but the employment that AAM generates are the ones that have integrated workforce pipeline development into their AAM planning from the beginning rather than treating it as a secondary consideration after the infrastructure decisions have been made.

The Strategic Takeaway: The Next Three Years Will Decide Everything

Advanced Air Mobility is not a distant future. It is a present-tense engineering, regulatory, and political project whose most consequential decisions are being made right now. The type certifications that Joby and Archer are pursuing will, when issued, establish the regulatory precedents that govern every eVTOL aircraft certified thereafter. The vertiport infrastructure investments being made today will determine the geographic shape of the first commercial AAM network. The workforce training programs being developed today will produce the pilots, technicians, and vertiport operators who will staff the industry's commercial launch, or fail to produce them, creating the same kind of talent gap that has constrained small UAS deployment for the past decade.

AAM is not waiting for the technology to catch up. The technology has largely arrived. What is being built now are the institutions, the infrastructure, and the workforce that will determine whether the technology serves the public or simply impresses it.

The competitive stakes of the next three years extend beyond the individual companies racing for type certification. The United States currently leads the global AAM certification race by the most credible available measures, Joby and Archer are ahead of their international competitors, the FAA's Part 21 certification framework is the global reference standard, and American AAM technology is being developed for international deployment as well as domestic operations. But the lead is not permanent. Competing certification programs in the United Kingdom, Europe, and Asia are advancing. International investors are funding non-American AAM companies with the same urgency that American venture capital brought to the sector. The infrastructure investments, workforce pipelines, and community acceptance frameworks that the United States builds over the next three years will either consolidate the competitive position its technology companies have established or allow it to erode while the industry's commercial geography is determined elsewhere.

Part IV began with an eight-minute flight. It ends with an argument about decades. The decisions being made right now, in certification offices, in city planning departments, in community college curriculum committees, and in the infrastructure boardrooms where vertiport development deals are being negotiated, will shape the urban sky for the rest of this century. The technology that makes those eight minutes possible already exists. The institutions, the workforce, and the political frameworks that make them routine are what the next phase of work must build.

Looking Ahead — Chapter Fourteen

Chapter 14 returns to the book's central argument: that the future of American leadership in uncrewed aviation will be determined not by the technology alone but by the decisions being made right now about workforce investment, industrial strategy, regulatory architecture, and the national will to build what the moment requires.

Notes and Sources

1. Joby Aviation. FAA Type Certification progress updates, 2025–2026. Stage 4 conformity review completed March 2026; Stage 5 compliance flight testing

underway; type certificate issuance and late 2026 commercial launch targeted in partnership with Delta Air Lines for New York City and Los Angeles. Sources: Aircraft Insider (April 19, 2026); Altitudes Magazine (April 2–5, 2026); Joby Aviation investor communications.

2. Archer Aviation. FAA Type Certification update, 2025–2026. 100% Method of Compliance acceptance; Midnight aircraft flight test expansion. Source for Archer certification status.

3. Kirsten Korosec. "Defunct electric aircraft startup Lilium's tech lives on over at Archer." TechCrunch, October 16, 2025. Lilium insolvency October 2024; Archer acquisition of 300 Lilium patents for €18 million. Source for Lilium cautionary case study.

4. Sterne Kessler. "Turbulence in the eVTOL IP Wars: Joby Attempts to Ground Archer." November 21, 2025. Archer 1,000+ patent portfolio; LA 2028 Olympics air taxi selection; Archer total funding exceeding $1.5 billion.

5. Mike Hirschberg. "eVTOL Leaders Find Uneven Success." Vertiflite, November/December 2024. Archer additional capital raises; Stellantis manufacturing commitment; United Airlines strategic investment.

6. Federal Aviation Administration. Advanced Air Mobility (AAM) Concept of Operations, Version 1.0. FAA, 2023. Defines the operational framework, airspace architecture, and infrastructure requirements for initial AAM operations.

7. NASA. Advanced Air Mobility Mission Overview. NASA.gov, accessed May 2026. NASA's role in AAM ecosystem development, including National Campaign flight demonstrations and UTM-AAM integration research.

8. Joby Aviation. Vertiport development partnerships and infrastructure planning disclosures, 2024–2026. Source for vertiport power infrastructure estimates: 2–5 MW early operations; 10–50 MW at scale.

9. Federal Aviation Administration. Noise Standards for Electric Vertical Takeoff and Landing Aircraft. FAA rulemaking notices, 2024–2025. Source for AAM noise certification framework discussion.

10. Wisk Aero. Autonomous certification pathway overview. Wisk.aero, accessed May 2026. Source for Wisk autonomous certification strategy discussion.

11. Federal Aviation Administration. Part 21 Type Certification process overview; Part 135 air carrier certificate requirements for AAM operations. FAA.gov. Primary regulatory framework source for certification and operations discussion throughout.

12. Archer Aviation. Anduril defense partnership announcement; 2028 Los Angeles Olympics air taxi provider selection. Multiple press releases, 2024–2026. Source for Archer multi-path revenue strategy.

13. Beta Technologies, Inc. (NYSE: BETA). SEC filings and investor communications, 2025–2026, including Q3 2025 earnings release (December 4, 2025) and Q1 2026 earnings release. Sources for: 85,000+ hours of flight and

ground testing; CX300 certification target late 2026 / early 2027; ALIA 250 eVTOL certification target approximately one year later; 7 of 8 FAA eVTOL Integration Pilot Program selections across 26 states; $300 million GE Aerospace strategic investment; UPS and United Therapeutics customer relationships. First U.S. passenger-carrying electric aircraft flight: June 2025. Robb Report (November 10, 2025) and Capital.com for IPO and certification timeline context.

Policy, Regulation, and the Airspace of Tomorrow

"The airspace of tomorrow is not waiting to be discovered. It is waiting to be decided, in rulemaking dockets, legislative hearings, and investment strategies, by the people and institutions willing to show up and do the work."

— Sharon Rossmark

The Negotiation Most Americans Will Never See

The comment period closes at 11:59 p.m. on a Tuesday in October. By mid-evening, the FAA's BVLOS docket has received approximately 3,100 public comments, a volume that, measured against the typical federal rulemaking, reflects genuine and concentrated stakeholder engagement.[3] Inside the FAA's Washington headquarters, staff attorneys and aviation safety engineers are already preparing for the months of analysis ahead. Outside, the industry has made its case: drone operators, manufacturers, agricultural associations, civil liberties organizations, public safety agencies, counter-UAS companies, and research universities have each submitted their arguments. What they have submitted is not merely opinion. It is the record on which the rule will be built, the evidentiary foundation that will determine, for years to come, what is legal, what is prohibited, and what is merely deferred.

This is where the future of uncrewed aviation is actually shaped, not in hangars or test ranges or investor pitch rooms, but in rulemaking dockets, congressional hearings, and interagency negotiations conducted in language that most Americans will never read and whose implications most of the industry's own workforce does not fully understand. Every operational freedom that a commercial drone operator exercises today exists because someone, at some point, engaged the regulatory process with sufficient expertise, persistence, and credibility to earn it. Every operational constraint that limits what the industry can do today exists for the same reason, or because the industry did not engage the process when the rule was being written.

This chapter is a guide to what is at stake in the ongoing regulatory negotiation, and to what the industry, its investors, its workforce developers, and its public champions must do to shape it. The preceding chapters have mapped the landscape, the talent gaps, the capital flows, the manufacturing vulnerabilities, the defense dependencies, and the regional competitive dynamics that define the uncrewed aviation ecosystem. This chapter asks the forward question: given all of that, what does a coherent national strategy for uncrewed aviation leadership actually look like, and who is responsible for building it?

The Regulatory Stack: A Decade of Layered Decisions

The regulatory framework governing UAS operations in the United States did not emerge from a single act of legislative foresight. It developed in layers, each one a distinct episode in the evolution of an industry that moved faster than the institutions governing it, requiring regulatory responses that were simultaneously overdue when they arrived and incomplete when they were finalized. Understanding the stack is a prerequisite for understanding the agenda that remains unfinished.

Part 107, which took effect on August 29, 2016, was the foundation: the first comprehensive legal framework for commercial small UAS operations, establishing certification requirements for remote pilots, defining the operational envelope within which commercial drone work could be legally conducted, and creating the waiver system through which operators could seek deviations from the standard rules for specific missions.[1] The 492,311 active Part 107 Remote Pilot Certificates recorded by the FAA as of December 31, 2025 are the accumulated product of a decade of professional development that Part 107 made possible.[4]

Remote Identification, the rule requiring drones to broadcast their identity, location, and operator location in real time, added the accountability layer that Part 107 had not addressed. Mandatory Remote ID compliance took effect on September 16, 2023, after years of rulemaking and industry debate about its technical implementation.[2] Remote ID is the data layer on which future airspace management systems depend. Without the ability to identify which aircraft is operating where, the UTM architecture that scales drone operations to commercial density is impossible to build safely.

UAS Traffic Management (UTM), developed through a multi-year NASA research program and advanced through the FAA's partnerships with commercial service providers, is the automated coordination architecture for low-altitude drone operations.[11] Traditional air traffic control—based on voice communication between human controllers and human pilots—cannot scale to the density of simultaneous drone operations expected under commercial BVLOS deployment. UTM introduces a distributed, service-based system in which operators and approved UAS Service Suppliers exchange data through application programming interfaces, enabling strategic deconfliction and operational coordination without continuous human intervention. The UTM ecosystem is now partially operational: as of early 2026, the FAA's Near-Term Approval Process (NTAP) has approved multiple UAS Service Suppliers, and a multi-partner cohort of service providers and operators is using the USS Network to support commercial UAS operations in several U.S. regions. Full national deployment, however, depends on the FAA's final BVLOS regulatory framework and sustained federal funding beyond the current authorization cycle.

The regulatory stack is not a bureaucratic artifact. It is the accumulated architecture of a decade of hard negotiation between an industry that moves fast and an agency whose mandate is to ensure that nothing in the sky kills anyone.

The BVLOS rulemaking, published as an NPRM on August 7, 2025, is the fourth and most consequential layer of the stack. It is the rule that determines whether the most economically transformative class of drone operations, long-range infrastructure inspection, large-scale precision agriculture, commercial package delivery, emergency response in remote areas, becomes routine and scalable or remains the domain of one-off waivers and case-by-case approvals.[3] The stakes are not abstract: the difference between a permissive BVLOS framework and a restrictive one is measured in billions of dollars of economic activity, in hundreds of thousands of jobs, and in whether the United States consolidates its position as the world's leading drone operating environment or cedes that position to regulatory regimes that move faster with less caution.

BVLOS: The Rule That Will Decide the Industry's Future

Beyond visual line of sight (BVLOS) operations are the operational capability that unlocks the full economic potential of commercial drones. The visual line of sight constraint that Part 107 imposes, requiring the remote pilot or a visual observer to maintain unaided visual contact with the aircraft throughout the flight, is a reasonable safety measure for a nascent industry operating in a regulatory environment without the infrastructure to support more autonomous operations. It is a ceiling on commercial ambition in a mature one. A single operator watching a single drone in a defined visual range can accomplish a bounded set of missions. A BVLOS-capable operator managing multiple aircraft on pre-planned autonomous routes, supported by detect-and-avoid technology and integrated with a UTM coordination system, can accomplish an order of magnitude more, at lower cost per mission, with better safety outcomes from the reduced human intervention that automation enables, and with the scalability that enterprise and logistics clients require.

The Part 108 NPRM proposes a performance-based framework that replaces the existing case-by-case waiver system with scalable, standardized rules for BVLOS operations. Industry responses to the NPRM reflected both strong support for the framework's direction and specific concerns about its implementation details: the treatment of existing Part 107 BVLOS waiver holders, the rule's emphasis on high automation over pilot-in-the-loop control, and the compliance burden its requirements would impose on small operators and public safety agencies that lack the organizational infrastructure of enterprise commercial operators.[7]

A June 2025 executive order directed the FAA to finalize the BVLOS rule within 240 days.[6] A 43-day government shutdown beginning October 2025 — the longest in U.S. history — extended the effective rulemaking timeline beyond the original February 2026 target; as of mid-2026 the final rule remains pending publication. The companies that are investing in BVLOS-ready technology and workforce ahead of the rule's finalization will be positioned to capture the first wave of commercial opportunity when the regulatory window opens. The companies waiting for the rule before making those investments will find themselves six to twelve months behind a competitive field that did not wait.

The workforce implications of BVLOS finalization deserve emphasis here because they have been addressed across multiple earlier chapters but bear direct repetition in the policy context. The Part 108 framework will require operators to demonstrate system competency at a level that exceeds what the existing Part 107 certification produces. The shift from one pilot managing one aircraft to one operator managing a fleet of autonomous vehicles, functioning more like an air traffic controller than a traditional drone pilot, is not merely a regulatory change. It is an occupational transformation that requires a new curriculum, a new credential structure, and a new training infrastructure that the education system is building in real time.

FAA Reauthorization and the Legislative Landscape

The FAA Reauthorization Act of 2024, Public Law 118-63, signed on May 16, 2024, was the most significant piece of aviation legislation in years, and its UAS-specific provisions established a legislative foundation on which the executive actions of 2025 were able to build.[5] Section 925 established a $6 million annual appropriation for the FAA's seven designated UAS test sites through fiscal year 2028, structured as matching funds for commercial technology demonstration and validation. The Act included mandates for BVLOS rulemaking progress, Remote ID enforcement timelines, UTM development requirements, and counter-UAS coordination directives that gave the industry a multi-year policy roadmap with legislative authority behind it.

But the legislative agenda for uncrewed aviation is far from complete, and the gaps in the current framework are consequential enough to deserve specific enumeration. Long-term funding for UTM deployment has not been secured beyond the current authorization cycle. Expanded counter-UAS authority for state and local law enforcement, airport operators, and critical infrastructure owners remains legally uncertain. A national drone infrastructure strategy, the coordinating mechanism that would align federal investment across the FAA, NASA, DoD, USDA, EDA, and DOL into a coherent industrial policy rather than a collection of parallel programs, does not exist in statutory form. Workforce development mandates that would direct Labor Department resources toward drone-specific training pipelines with the clarity and urgency that the talent shortage demands remain aspirational rather than funded.

The next FAA Reauthorization cycle, anticipated in 2028, represents the most significant near-term legislative opportunity to address these gaps. The industry, its workforce development partners, and the regional economic development communities that have committed to drone economy leadership have a shared interest in engaging that process early, specifically, and with the technical credibility that translates stakeholder interests into statutory language. Reauthorization hearings begin years before the legislation is drafted. The record that shapes the final bill is built in testimony, comment letters, and technical briefings that begin now.

Remote ID: The Digital Backbone of Drone Accountability

Remote Identification is the regulatory achievement that is most frequently underestimated in its strategic significance. Its public characterization as a digital license plate for drones is accurate but incomplete. A license plate tells an observer who owns a vehicle. Remote ID tells the airspace management system, law enforcement, counter-UAS operators, and the UTM service provider network where every broadcasting aircraft is, where its operator is located, and whether its identity matches an authorized flight. That information, aggregated across all operating aircraft in a given airspace, is the situational awareness layer on which every subsequent operational capability depends.

The compliance landscape for Remote ID is now more complete than enforcement resources have been able to fully monitor. The FAA's mandatory compliance date of September 16, 2023 marked the point at which new drone sales in the United States were required to include Remote ID broadcast capability, and at which operators were required to fly only Remote ID-compliant aircraft or within FAA-Recognized Identification Areas (FRIAs). The transition has been substantially complete for commercially manufactured aircraft, where manufacturers had strong incentives to integrate compliance into new products. The retrofit market, the large installed base of pre-compliance aircraft, remains a persistent compliance gap that the FAA's enforcement resources have not fully addressed.

The more important story about Remote ID is not its compliance rate. It is its enabling function. The BVLOS regulatory framework that Part 108 is building assumes Remote ID as a foundational input: a BVLOS operator's aircraft must be broadcasting its identity and position for the UTM coordination system to manage its interaction with other aircraft in the shared airspace. Remote ID is not the end of drone accountability. It is the beginning of the automated accountability architecture that every subsequent operational ambition requires.

Counter-UAS: The Parallel Policy Crisis

The expansion of legitimate drone operations has been accompanied, with uncomfortable symmetry, by the expansion of illegitimate ones. Drones near airports, over public events, around critical infrastructure, and in sensitive airspace represent a threat category that has grown faster than the legal framework governing responses to it. The incidents are not hypothetical: runway incursions, stadium overflight events, prison contraband delivery, wildfire reconnaissance interference, and industrial espionage attempts have all been documented at a frequency that has moved counter-UAS from a niche security concern to a mainstream public safety and national security issue.

The legal constraint that compounds this operational problem is among the most consequential policy gaps in the current framework. Federal law restricts the authority to detect, track, disrupt, or disable a drone in flight to a specific set of federal agencies, primarily the Department of Defense, the Department of Homeland Security, and the Department of Justice under the 2018 FAA Reauthorization Act's counter-UAS provisions.[8] State and local law enforcement agencies, airport authorities, stadium operators, and critical infrastructure owners have limited legal authority to take action against an unauthorized drone even when the threat is unambiguous and the window for response is measured in seconds. The gap between the operational reality that local public safety agencies face and the legal authority they are permitted to exercise has created a dysfunction that incidents at airports, major sporting events, and critical facilities have made impossible to ignore.

Counter-UAS is no longer a niche issue. It is a national security imperative. And the policy framework for addressing it is at least three years behind the operational reality it needs to govern.

The policy challenge is to expand counter-UAS authority in a way that is operationally effective without creating a legal architecture that enables the misuse of drone interdiction technology against lawful operations. The technologies capable of disabling a rogue drone, radio frequency jamming, GPS spoofing, directed energy systems, are not surgically precise in their effects. A jamming system deployed at an airport to disable an unauthorized drone in the approach corridor may simultaneously disrupt the Remote ID broadcasts of legitimate aircraft operating in the area. These collateral effects require that counter-UAS authority be exercised with technical sophistication and legal accountability that general-purpose law enforcement agencies cannot currently be presumed to possess.

The federal government's current approach, piloting expanded counter-UAS programs through the Department of Homeland Security and the FAA at specific high-risk venues while developing the statutory and technical framework for broader authority expansion, is producing the evidence base that a responsible legislative solution requires. The next FAA Reauthorization cycle is the most likely vehicle for a comprehensive counter-UAS authority framework that extends meaningful interdiction capability to state and local actors within a structure of technical standards, training requirements, and accountability mechanisms that protects legitimate operations while enabling genuine threat response.

State and Local Preemption: The Patchwork Problem

The FAA's exclusive authority over the navigable airspace of the United States is a well-established legal principle, affirmed in statute and in federal court decisions. In practice, its application to the complex operational environment of commercial drone operations has generated a persistent tension between federal preemption and legitimate local governance interests that the current legal framework has not resolved with the clarity that the industry requires.

State legislatures have enacted drone-specific legislation governing privacy, surveillance, law enforcement use, agricultural operations, and critical infrastructure protection, statutes that reflect genuine constituent concerns about the implications of pervasive aerial surveillance and that exercise legal authorities that states have traditionally held over land use and privacy. Cities have enacted ordinances governing where drones can operate over public parks, what distances aircraft must maintain from residential properties, and what noise standards apply to commercial drone operations in residential zones. These enactments are not, in most cases, the product of regulatory hostility toward the drone industry. They are the product of elected officials responding to constituents who have legitimate interests in how the airspace above their communities is used.

The risk is not that any single state or local law creates an insurmountable operational barrier. It is that the accumulation of inconsistent requirements across jurisdictions creates the compliance complexity, legal uncertainty, and operational fragmentation that makes national-scale commercial drone services economically unviable. A delivery service planning a network of BVLOS routes across three states cannot afford to navigate a different regulatory regime in each municipality along the route. An infrastructure inspection company whose utility corridor contracts span state lines cannot afford the legal exposure of operating under rules whose preemption status has not been authoritatively resolved.

The industry's interest in a clear preemption framework, one that reserves federal authority over airspace and operational safety while providing state and local governments with defined roles in land use decisions, privacy protection, and community impact mitigation, aligns with the interests of the local governments that want regulatory clarity as much as the operators do. The current ambiguity serves no one well except the attorneys litigating its resolution. A legislative solution that defines the boundaries of federal preemption with precision, establishes a framework for state and local coordination with the FAA on community impact issues, and creates a dispute resolution mechanism for conflicts between operators and local authorities would reduce the legal overhead that currently falls on every party in the system.

Industry Advocacy: A Strategic Capability, Not a Side Function

The drone industry's ability to influence the policy environment that governs its operations is uneven in ways that correlate directly with organizational scale and regulatory sophistication. The largest manufacturers, the best-capitalized operators, and the most established industry associations engage the FAA rulemaking process, the congressional staff responsible for aviation legislation, and the interagency working groups that coordinate federal UAS policy with the depth and consistency that effective policy engagement requires. The majority of the industry, the mid-market service companies, the regional operators, the community college training programs, the drone agriculture specialists, the public safety agencies building drone programs on constrained budgets, participate sporadically if at all, arriving at public comment periods without the regulatory fluency, the technical documentation, or the organizational continuity to make their perspectives as legible to rulemakers as the perspectives of the large, well-resourced stakeholders who appear at every hearing.

The organizations doing the heaviest lifting in drone industry policy engagement include the Association for Uncrewed Vehicle Systems International (AUVSI), the Small UAV Coalition, the Commercial Drone Alliance, the Drone Service Providers Alliance (DSPA), which specifically represents the small commercial drone service businesses that constitute the majority of the operational industry, and the state-level UAS councils that have formed around FAA test sites and regional economic development initiatives.[10] These organizations represent the policy infrastructure through which the industry's collective interests are translated into regulatory comment letters, legislative testimony, and the sustained relationships with agency staff and congressional offices that make institutional knowledge transferable across election cycles and personnel transitions.

Effective policy engagement is not a reactive function, monitoring rules as they are proposed and submitting comments before they close. It is a proactive one: shaping the agenda that determines which rules get written, in what sequence, with what framing, and toward what operational outcomes. The companies and organizations that understand the FAA's internal planning process, that participate in the aviation rulemaking committees and standards bodies where regulatory concepts are developed before they reach the NPRM stage, and that invest in the technical credibility and long-term relationship capital that regulatory influence requires are operating at a fundamentally different level of strategic engagement than those that treat policy as a compliance function.

What a National Strategy for Uncrewed Aviation Leadership Actually Requires

Fourteen chapters of argument converge on a single claim: the United States has the technology, the capital, the regulatory foundations, and the industrial base to lead the world in uncrewed aviation for the next generation. What it does not yet have is the coordinated national strategy that would convert those assets into durable leadership. The gaps, in workforce, in domestic manufacturing capacity, in investment alignment, in regulatory coherence, and in the institutional architecture that connects federal investment to local implementation, are documented throughout this book and are known to the policymakers, industry leaders, and workforce developers who are working to close them. The question is not whether these gaps are real. It is whether the institutional will exists to address them at the scale and with the coordination they require.

A genuine national strategy for uncrewed aviation leadership requires five things that the current policy environment provides only partially and inconsistently. The first is a coordinating architecture, the interagency mechanism that aligns the FAA's regulatory investment, NASA's foundational research, DoD's procurement commitments, the Department of Labor's workforce development funding, the USDA's rural application programs, and the EDA's regional economic development resources around a shared geographic and industrial strategy. The absence of this architecture, documented in Chapter 9 through the experience of the regional airport director whose inbox contained four uncoordinated federal program solicitations on a single Monday morning, is the single most important structural deficiency in the current federal approach to the drone industry.

The second requirement is a funded, sustained workforce development infrastructure that is calibrated to the industry's actual talent needs rather than to the existing categories of federal workforce investment. A national strategy that treats workforce development as a first-order industrial policy priority, rather than as a downstream consequence of technology investment, would align the WIOA framework, the FAA's UAS Collegiate Training Initiative (CTI) program, and the Labor Department's apprenticeship system, directing them toward the specific credential pathways that employers value and investing in the community college and technical school infrastructure that can reach the full depth of the available talent pool.

The third requirement is a domestic manufacturing policy that extends beyond procurement restrictions to address the component supply chain vulnerabilities that NDAA compliance alone cannot solve.[9] A supply chain that achieves NDAA compliance at the finished-product level while remaining dependent on Chinese-processed rare earth minerals, Chinese-manufactured battery cells, and Chinese-produced permanent magnets has achieved regulatory compliance without supply chain sovereignty.[12] The manufacturing policy agenda that the next phase of American drone industrial strategy requires goes component by component through the supply chain and asks, for each critical input: what would it take to produce this domestically or from allied-nation sources at competitive cost and quality? That question must also extend to the consumer and prosumer market segment, the DJI-class drones used daily by small businesses, educators, workforce developers, and the community organizations building the talent pipelines the industry needs. The current manufacturing policy addresses the defense and enterprise procurement tier with NDAA compliance frameworks and Blue UAS certification. It does not yet address the supply chain for the affordable, accessible hardware that small commercial operators, workforce training programs, and educational institutions depend on, and for which no domestic alternative exists at a competitive price point. A complete domestic manufacturing strategy must account for this segment, not only the high-end defense procurement tier.

The fourth requirement is a regulatory modernization agenda that matches the pace of the industry's development. The BVLOS rulemaking is the most urgent item on this agenda, and its finalization will represent the most significant regulatory advance for the commercial industry since Part 107. But the regulatory agenda does not end with BVLOS. The counter-UAS authority framework, the state and local preemption resolution, the AAM certification precedents being established by Joby and Archer's type certification processes, the airspace integration architecture for combined UAS and AAM operations, each represents a regulatory decision point that will shape the operational environment for a decade or more.

The fifth requirement is an inclusive talent strategy that is explicitly designed to reach the full depth of the available workforce pool. The drone industry cannot fill its talent gap by recruiting from the same narrow pool that every other technology sector is already competing over. The mid-career professionals making deliberate pivots into drone careers, the veterans whose military UAS experience maps directly onto commercial requirements, the community college students in agricultural states who are closer to the industry's most economically significant applications than to any coastal technology hub, these populations represent the expansion of the talent base that the industry's growth requires. Reaching them requires visibility, mentorship, and economic framing that makes the opportunity legible to people who have no existing connection to aviation.

The Answer Is Not Yet Written

This book began with a Tuesday morning in Assumption, Illinois, a drone climbing into the pre-dawn dark above 200 acres of corn that needed spraying before the heat of the day set in. It moved through the Pittsburgh power corridor, the Phoenix doorstep, the Gulf of Mexico, and the Texas borderland. It traced the regulatory arc from the legal gray zone of pre-Part 107 commercial operations to the threshold of the BVLOS era. It followed the supply chain vulnerabilities from Skydio's battery crisis to the permanent magnet dominance that lies beneath every drone motor in the sky. It mapped the talent gaps from the unfilled BVLOS operations specialist positions to the missing drone technician credential that has no federal standard. It documented the capital flows from the SPAC-era peak to the disciplined strategic investment of 2025 and 2026. It argued the regional competition from North Dakota's decade of patient infrastructure investment to the community that is just now lacing its shoes. And it arrived here: at the policy chapter, at the regulatory negotiation, at the place where all of it either comes together or falls apart.

The technology is not the bottleneck. The market is not the bottleneck. The regulatory path is clearing. The workforce, the investment architecture, and the manufacturing base are the work still to be done, and that work is within the country's own control.

The argument of this book can be stated plainly: the United States has a generational opportunity to lead in one of the defining technology sectors of the 21st century. The technology is largely in place. The regulatory foundation, while incomplete, is being built with more seriousness and speed than the history of comparable technology sectors might have predicted. The capital is flowing. The demand is real. What stands between today's momentum and the full realization of uncrewed aviation's potential is a set of human and institutional challenges that are, of all the constraints the industry faces, the ones most directly within the country's own control. Capital flows where it will. Adversary nations manage their own supply chains. But the United States can choose, as a matter of deliberate national policy and coordinated industry investment, who it trains, how it trains them, where it builds, how it coordinates, and how it regulates. Those choices, made well, compound over a decade into durable industrial leadership. Made poorly, or not made at all, they compound into a different kind of lesson.

The next decade will answer the question this book has been asking. The answer is not yet written. But the people writing it are identifiable: they are the workforce developers building credential pathways in communities where the aviation industry has no prior presence; the investors allocating patient capital toward the domestic manufacturing infrastructure that procurement policy has demanded and the market alone has not built; the regulatory affairs professionals showing up at every rulemaking comment period with the technical depth to make the industry's operational reality legible to the people writing the rules; the community college instructors teaching drone maintenance to students who will spend their careers maintaining aircraft that do not yet exist; the regional economic development directors leading with workforce in their site selection pitches and winning; and the decision-makers in every institution that touches this industry, corporate, governmental, educational, and philanthropic, who understand that the airspace of tomorrow is not waiting to be discovered. It is waiting to be decided.

Looking Ahead — Chapter Fifteen

The regulatory and policy architecture described in this chapter is the foundation on which the integrated strategy must be built. Chapter 15 takes the workforce argument, the investment argument, and the manufacturing argument developed across this book and assembles them into the coordinating architecture that all three together require: the aligned investment targets and accountability mechanisms that would convert the United States' genuine competitive assets into durable industrial leadership.

Notes and Sources

1. Federal Aviation Administration. Part 107 — Small Unmanned Aircraft Systems; Final Rule. 14 CFR Part 107. Effective August 29, 2016. Primary statutory framework for commercial UAS operations.

2. Federal Aviation Administration. Remote Identification of Unmanned Aircraft; Final Rule. 86 FR 4390. Mandatory compliance date: September 16, 2023.

3. Federal Aviation Administration. Notice of Proposed Rulemaking: Normalizing Unmanned Aircraft Systems Beyond Visual Line of Sight Operations (Part 108). 90 FR 38212. Docket No. FAA-2025-1908. Published August 7, 2025. Comment period closed October 6, 2025. Approximately 3,100 public comments filed.

4. Federal Aviation Administration. U.S. Civil Airmen Statistics, December 31, 2025. FAA Aeronautical Center, Oklahoma City, OK. Active Part 107 Remote Pilot Certificates: 492,311.

5. FAA Reauthorization Act of 2024, Public Law 118-63. Signed May 16, 2024. UAS-specific provisions including Section 925 test site funding, BVLOS rulemaking mandates, Remote ID enforcement, UTM development requirements, and counter-UAS coordination.

6. White House. Executive Order 14307: "Unleashing American Drone Dominance." Signed June 6, 2025. Directed FAA to issue BVLOS proposed rule within 30 days; 240-day finalization directive; prioritization of U.S.-manufactured UAS.

7. DLA Piper. "FAA's Proposed Part 108 BVLOS Rule: Industry Response and Key Concerns." October 13, 2025. Summary of stakeholder comments including waiver holder treatment, automation requirements, and small operator burden.

8. FAA Reauthorization Act of 2018, Public Law 115-254. Section 1602 counter-UAS authority provisions for DHS, DOD, and DOJ.

9. National Defense Authorization Act for Fiscal Year 2024, Public Law 118-31. American Security Drone Act; FAR interim rule effective November 12, 2024.

10. Association for Uncrewed Vehicle Systems International (AUVSI). Industry policy positions and regulatory engagement documentation, 2024–2026. Also:

Drone Service Providers Alliance (DSPA). dspalliance.org. Represents small commercial drone service providers in federal policy engagement; primary advocacy organization for the small business segment of the commercial UAS industry.

11. Multiple sources. (a) NASA Aeronautics Research Mission Directorate. "UAS Traffic Management (UTM) Project." NASA.gov. Project concluded May 2021. Describes UTM as a multi-year research initiative to enable safe, scalable low-altitude UAS operations through a distributed, service-based architecture; research results transferred incrementally to the FAA beginning 2020. Available: nasa.gov/directorates/armd/past-armd-projects/utm-project/. (b) Federal Aviation Administration. Unmanned Aircraft System (UAS) Traffic Management (UTM) Concept of Operations, Version 2.0. FAA Office of NextGen, Washington, D.C. March 2, 2020. Establishes that traditional ATC cannot scale to the anticipated density of UAS operations and that UTM is designed to operate separate from, but complementary to, ATC; defines the federated, API-based architecture through which operators and UAS Service Suppliers (USS) exchange operational data. Available: faa.gov/sites/faa.gov/files/2022-08/UTM_ConOps_v2.pdf. (c) Federal Aviation Administration. UAS Traffic Management (UTM) — Near-Term Approval Process (NTAP). FAA UAS Integration Office, 2024–2026. FAA issued its first NTAP Letter of Acceptance (LOA) in May 2024; multiple USS providers including Airspace Link, OneSky, and NUAIR have received LOAs through 2025–2026 for strategic deconfliction services supporting commercial BVLOS operations. Available: faa.gov/uas/advanced_operations/traffic_management. (d) U.S. Government Accountability Office. Drones: FAA Should Improve Its Approach to Integrating Drones into the National Airspace System. GAO-23-105189. Published January 26, 2023. Finds that FAA lacks a comprehensive drone integration strategy and that national UTM deployment depends on BVLOS rulemaking progress and a sustained implementation framework. Available: gao.gov/products/gao-23-105189. (e) U.S. Department of Transportation, Office of Inspector General. FAA Has Made Progress in Advancing BVLOS Drone Operations. OIG Report. June 30, 2025. Confirms that national UTM deployment remains contingent on FAA's BVLOS rulemaking finalization and sustained federal funding for UTM infrastructure and integration.

12. One Big Beautiful Bill Act (OBBBA), signed July 4, 2025. $1.4 billion appropriation for small UAS industrial base.

The Integrated Strategy: Workforce, Investment, and Manufacturing

Why Three Separate Conversations Must Become One

"No great industrial sector was built by a workforce program, an investment strategy, or a manufacturing policy working alone. Every one of them was built by all three working together, deliberately, in sequence, with each element reinforcing the others. The countries that understand this are building right now. The only question is whether we are building with them or watching from the hallway."

— Sharon Rossmark

Three Rooms, One Industry, and the Missing Conversation

The conference center hallway connects three rooms. In Room A, the workforce development panel is underway: training directors from community colleges, FAA CTI program coordinators, and military transition specialists discussing credential pipelines and technician certification gaps. In Room B, the investment and venture capital panel is running through the capital cycle: defense primes investing in autonomy startups, sovereign wealth funds backing domestic manufacturing, series rounds at valuations that would have been unthinkable four years ago. In Room C, the domestic manufacturing panel is deep in supply chain specifics, battery cell sovereignty, rare earth processing dependencies, the precision machining workforce that drone production at scale will require.

All three sessions are running simultaneously. The attendees in each room are largely different people. No one is moving between them. The workforce developers are not talking to the manufacturers about what skills the factory floor will require in 24 months. The investors are not talking to the workforce developers about which training programs have the credibility to build the talent the companies they are funding will need. The manufacturers are not talking to the investors about which component supply chain investments would produce the greatest reduction in strategic vulnerability per dollar deployed.

In the hallway, a senior official from a foreign ministry, representing a country that has been running all three conversations in a single room for the past five years, pauses, looks at the three separate doors, and makes a note in her phone. She is not surprised. She has seen this pattern before in other American technology sectors, at other inflection points, and she knows how it tends to resolve: either someone builds the bridge between the rooms before the competitive window closes, or the rooms remain separate until the window has closed and the post-mortem begins. The note, if we could read it, would say something like: the pieces are all there. The architecture connecting them is not.

This chapter is about what it would take to build that architecture. Not as a theoretical exercise, but as a practical agenda: the specific investments, institutional structures, and coordinating mechanisms that would allow the United States' genuine advantages in workforce capability, capital depth, and manufacturing potential to compound into the durable industrial leadership that the individual advantages, in isolation, cannot produce.

Why Integration Matters: The Systems-Level Argument

Every chapter of this book has documented a version of the same structural problem from a different vantage point. Chapter 3 found that the talent deficit was not merely a pipeline problem but a systemic one: the industry cannot fill its workforce gap without the investment to fund training programs, and training programs cannot attract the investment they need without the market clarity that comes from a functioning domestic manufacturing sector producing platforms that buyers are purchasing at scale. Chapter 9 found that federal investment in uncrewed aviation is real and substantial but fragmented across agencies that do not coordinate with each other in ways that compound rather than dissipate its effects. Chapter 10 found that private capital flows toward hardware and software because they produce the returns that venture investors are structured to chase, while the workforce infrastructure, domestic manufacturing capacity, and regional ecosystem development that would make the industry's growth sustainable attract investment only when someone has explicitly designed the mechanism through which returns flow back to investors.

The pattern across all three observations is the same: each element of the industry's development, workforce, investment, manufacturing, is necessary but not sufficient on its own. A trained workforce without companies hiring at scale, equipment to operate, and regulatory frameworks enabling the operations those companies need is a workforce waiting for an industry that has not yet arrived at the scale required to employ it. A domestic manufacturing base without the skilled technicians to staff the factories, the engineers to improve the designs, and the operators to demonstrate the products' value to enterprise customers is a stranded asset. Capital deployment without coordination fills some gaps, misses others, funds duplicative efforts, and systematically under-invests in the categories, workforce training infrastructure, domestic component supply chains, regional ecosystem development, where the returns are real but the timelines are longer than venture capital structures are typically designed to accommodate.

Workforce, investment, and manufacturing are not three separate pillars of the drone economy. They are one system. If any element is missing, the others cannot compensate. The product of their multiplication is national capability, and zero times anything is zero.

The systems-level insight is not a novel one in the history of American industrial policy. The Interstate Highway System required the simultaneous coordination of federal funding, state implementation capacity, engineering workforce, construction industry scale, and a legal framework for land acquisition that no single element of the system could have produced alone. The CHIPS Act's semiconductor manufacturing revival requires the simultaneous availability of federal incentive capital, private manufacturing investment, engineering talent, supply chain partnerships, and the regulatory certainty that makes decades-long facility investment commercially rational.[1] In each case, the insight that made coordinated industrial strategy necessary was the same: the individual elements of capability are insufficient without the architecture that connects them into a functioning system.

China: The Most Compelling and Uncomfortable Comparison

China's dominance in commercial drone manufacturing, which produced DJI's extraordinary global market position and the strategic vulnerabilities that the United States is now urgently trying to address through NDAA procurement restrictions, the Blue UAS framework, and the OBBBA's $1.4 billion industrial base investment, was not an accident of market forces.[2,3] It was the result of a deliberate, integrated national strategy executed over more than a decade with the consistency of purpose and the institutional patience that the U.S. drone industry's fragmented development has not matched.

China's system succeeded because it operated as a system. State investment in manufacturing capability created the production infrastructure that commercial success required. Educational policy aligned university engineering programs and vocational training toward the technical skills that drone production needed, producing a workforce that was available when the manufacturing infrastructure needed it rather than years behind it. Domestic market preference through government procurement policy provided the guaranteed demand that justified manufacturing investment before the export market was large enough to justify it on its own. Export market development through aggressive pricing, global distribution partnerships, and the strategic suppression of component costs built the market share that became a self-reinforcing competitive advantage.

This is not an argument for replicating China's model. The United States cannot and should not organize its industrial policy through the mechanisms of state-directed capitalism. It is an argument for understanding why the integrated approach succeeded, so that the United States can design its own version of that integration, one built on the mechanisms available in a democratic market economy: public-private investment vehicles, coordinated federal funding, aligned procurement standards, and the kind of patient institutional coordination that the CHIPS Program Office demonstrated is achievable within American governance structures when the strategic will to achieve it exists.

Allied Models: Proof That Integration Is Achievable

The most useful counter-evidence to the claim that integrated national drone strategy is achievable only in authoritarian governance systems comes from the United States' closest allies, democratic market economies that have developed coherent, cross-functional UAS strategies with measurable results.

Israel's defense UAS sector is among the most technically advanced in the world, and its competitive position derives directly from the integration of military procurement, manufacturing capacity, and workforce development that the country's security imperatives have enforced. Military procurement drives technology development with the urgency and performance standards that no commercial market alone could impose. Manufacturing capability grows alongside defense demand because the procurement framework requires domestic production. Veterans of the Israel Defense Forces' UAS units transition into a commercial sector that was built to employ them, carrying operational experience, technical training, and institutional knowledge that the commercial market values highly.

Integration is not a feature of authoritarian industrial policy. It is a feature of serious industrial policy, and the United States has demonstrated it can build seriously when the strategic will exists to do so.

Australia's national drone strategy, while more recently developed, offers a different model: a democratic middle-power country explicitly building a coherent UAS framework that connects regulatory development, training standards, domestic industry support, and airspace infrastructure investment through a single strategic architecture.[7] The United Kingdom's Future Flight Challenge, a £300 million program jointly funded by government and industry, has explicitly connected airspace integration work, vehicle development, and the demonstration of commercial operations in ways designed to produce regulatory evidence and workforce development in parallel with technology development.[6] The United States does not lack the resources to build a comparably integrated program. It has thus far lacked the institutional architecture to organize one.

The U.S. Coordination Gap: A Structural Problem With a Structural Solution

The United States enters this competition with extraordinary assets. Its research universities produce the engineering talent that autonomous systems development requires at a volume and quality that no other country matches. Its venture capital ecosystem has funded the most technically ambitious drone and eVTOL companies in the world. Its domestic manufacturing base has the industrial culture, the workforce foundation, and the supplier infrastructure to scale if the demand signal and the investment are sustained. Its regulatory safety culture, embodied in the FAA's rigorous but evolving approach to UAS integration, produces the trust infrastructure that commercial aviation requires to operate at scale in shared airspace.

What the United States currently lacks is the coordinating architecture that would allow these assets to compound rather than operate in parallel. The FAA manages airspace and certification. NASA conducts foundational research on UTM and advanced air mobility. The Department of Defense drives procurement and sets the security standards that shape the domestic manufacturing base. The Department of Labor administers the workforce development funding that community college drone programs depend on. The USDA funds rural UAS applications. The Economic Development Administration supports regional drone economy development. Each of these agencies has genuine and substantial engagement with the UAS sector. None of them has a statutory mandate to coordinate with the others.

The consequence is not malice or incompetence. It is the ordinary dysfunction of a government organized for functional specialization rather than systemic coordination, a government that is individually competent in each domain and collectively incoherent across them. The regional airport director whose Monday morning inbox contained four uncoordinated federal program solicitations, described in Chapter 9, is experiencing the cumulative effect of eight agencies each executing their mandates efficiently and in isolation. The cost of that isolation is borne by the local actors who must navigate it, the taxpayers whose investments are not compounding, and the national competitive position that is not being built as deliberately as the strategic moment requires.

What an Integrated National Strategy Would Actually Look Like

An integrated national UAS strategy does not require the creation of a new federal department or a reorganization of existing agencies. It requires three things: a coordinating architecture, aligned investment targets, and accountability mechanisms. The CHIPS Program Office model demonstrates that the U.S. government can build effective industrial strategy coordination within existing institutional structures when the strategic will to achieve it exists.[1] A UAS equivalent would require comparable design but could achieve comparable results.

The workforce pillar of an integrated strategy begins with establishing a national credential standard for UAS maintenance technicians, the regulatory baseline that does not yet exist and whose absence is documented in Chapter 3. Alongside the credential standard, a funded expansion of the FAA's UAS Collegiate Training Initiative network, from its current 145-plus institutions toward the 300-plus that the BVLOS era's workforce demand will require, needs to be accompanied by the employer commitment mechanisms that convert training programs into actual pipelines rather than credential-production exercises disconnected from hiring decisions.[4,5] Scholarship programs for workers entering UAS careers from non-traditional pathways, and a systematic connection between the separation process for UAS-trained service members and the commercial employers who need exactly the skills those service members carry, would expand the talent base beyond the narrow pool that the industry has historically recruited from.

The manufacturing pillar requires a level of supply chain specificity that current policy does not provide. The NDAA procurement restrictions and the Blue UAS framework address the finished-product layer of the dependency problem.[3] The component supply chain, battery cells, rare earth processing, permanent magnets, precision sensors, requires a different and more granular policy response: industrial development investment targeted at the specific chokepoints that the Skydio battery crisis of October 2024 identified, structured as public-private co-investment through Defense Production Act authorities and industrial base investment mechanisms that the semiconductor supply chain strategy has employed.[10]

The investment alignment pillar is the one most dependent on institutional design rather than direct spending. Coordinated federal research and development funding, aligned across FAA, NASA, DoD, and USDA so that each agency's research investments build toward shared technology readiness targets rather than agency-specific program metrics, would reduce duplication and accelerate the translation of research findings into deployable operational capability. Public-private investment vehicles, analogous to the energy innovation hubs that the Department of Energy has used to bridge federal research investment and private commercialization capital, could direct patient equity toward the domestic component manufacturing and regional infrastructure investments that venture capital is structurally unlikely to fund on the timelines and at the scale the strategic need requires.

Associations as Bridges: The Interim Architecture

In the absence of a formal federal coordination mechanism, the work of connecting the three conference rooms has fallen, by default, to the industry associations, academic consortia, and state-level UAS councils that have built the institutional relationships and technical credibility to function as de facto integrators. AUVSI's annual XPONENTIAL conference is, among other things, a physical version of the integrated strategy that the three-room problem describes: a single venue in which workforce developers, investors, manufacturers, regulators, and operators encounter each other's priorities in ways that the separate institutional silos of their normal operating environments do not produce.[12] The cross-pollination that occurs there is genuine and consequential. It is also voluntary, episodic, and dependent on the organizational bandwidth of individual participants rather than on institutional structures that would produce the same cross-pollination on a continuous basis. The Drone Service Providers Alliance (DSPA) represents a different but equally important piece of the bridge-building function: where AUVSI and similar organizations primarily serve manufacturers, enterprise operators, and large commercial stakeholders, the DSPA specifically advocates for the small commercial drone service businesses, the Part 107 operators running inspection, agriculture, media, and real estate services, that constitute the majority of the industry's actual operational workforce. An integrated national strategy that does not account for the needs, constraints, and policy interests of this segment is not a complete strategy. The DSPA gives this segment a policy voice that was previously absent from the organized advocacy landscape.

State-level UAS councils, the organizations that have formed around FAA test sites and regional drone economy initiatives in North Dakota, Virginia, Nevada, Texas, and a growing number of other states, represent a different expression of the integration function. North Dakota's UAS Council, which played a central role in the strategic sequencing of investments that produced the Vantis-Pathfinder integration and the $400 million leverage ratio documented in Chapter 9, is the most fully realized example of this function.[11] Its model is replicable, though it requires the institutional commitment and the long-term planning horizon that state governments have historically been reluctant to sustain across election cycles.

The limitation that all of these organizations share is the one that distinguishes interim coordination from genuine institutional architecture: they can build bridges between the rooms, but they cannot require anyone to walk across them. An industry association can convene a conversation between a workforce developer and a manufacturer, but it cannot compel the manufacturer to commit to hiring the developer's graduates. The voluntary coordination that associations and councils provide is valuable, necessary, and insufficient. It is the scaffolding for a national strategy, not the strategy itself.

The Strategic Window: Why the Next Two to Three Years Are Decisive

The competitive landscape in uncrewed aviation is not yet fixed. The manufacturing positions, the workforce pipelines, the airspace management architectures, and the regulatory frameworks that will define the industry's structure for the next generation are being established right now. The BVLOS rulemaking finalization, expected in late 2026 or early 2027, is the single most consequential near-term regulatory event in the commercial drone sector's history. When Part 108 takes effect, it will immediately create demand for BVLOS-capable operators, BVLOS-ready aircraft, UTM-integrated operations centers, and the insurance and liability frameworks that commercial BVLOS operations require. The companies and regions that have prepared will capture that demand. The ones that have not will spend twelve to eighteen months catching up to competitors who understood that preparation begins before the window opens, not after.

The international competitive pressure adds urgency to every domestic timeline. China's response to NDAA restrictions and Blue UAS requirements has not been retreat. It has been acceleration: continued investment in manufacturing capability, supply chain depth, and international market penetration in the markets where American procurement restrictions do not apply. European competitors are advancing their own certification frameworks, their own corridor infrastructure, and their own workforce development programs with the explicit goal of capturing the commercial AAM and advanced UAS market share that American regulatory constraints or institutional inaction might leave available. The window during which integrated national strategy produces durable competitive advantage rather than merely competitive recovery narrows every quarter that the three-room problem remains unsolved.

The Strategic Takeaway: Build the System

The United States does not lack talent. It does not lack capital. It does not lack manufacturing capability. It does not lack the innovation culture that produces the technology breakthroughs the industry requires. What it currently lacks is the institutional architecture that allows those assets to function as a system rather than as a collection of impressive but uncoordinated elements operating in parallel.

Building the system is not a single policy decision or a single legislative act. It is a sequence of decisions, each of which creates the conditions for the next. A national UAS credential standard for maintenance technicians creates the hiring clarity that drives employer commitment to training programs. Employer commitment to training programs creates the pipeline evidence that justifies community college investment in UAS curriculum. Community college UAS curriculum, scaled across the 300-plus institutions that the BVLOS era's workforce demand requires, creates the talent availability that makes domestic manufacturing investment in workforce-dependent regions commercially rational. Domestic manufacturing investment in workforce-dependent regions creates the supply chain depth that reduces the component-level vulnerabilities that NDAA compliance at the finished-product level alone cannot address.

The individual pieces of American drone industry leadership already exist. The missing piece is the architecture that connects them, the bridge between the rooms where the workforce developers, the investors, and the manufacturers are each working hard in isolation.

The sequence is not automatic. It requires institutional decisions at each step: the federal standard-setting that gives credential pathways the market authority they need to drive employer behavior; the investment structures that direct capital toward the components of the system that markets underfund; the procurement frameworks that provide the demand certainty that manufacturing investment requires; and the governance mechanisms that hold each element of the system accountable to the shared targets that make the whole greater than the sum of its parts. None of these decisions is technically difficult. Each of them requires the political will, the institutional patience, and the long-term thinking about competitive advantage that short-term optimization tends to crowd out.

The foreign ministry official in the hallway made her note and moved on. The answer to what she wrote down is not a press release or a policy white paper. It is the decision, made by the right people with the right authority, to open all three doors at once and hold the conversation that the industry, the country, and the moment require.

Looking Ahead — Chapter Sixteen

Chapter 16, the final chapter of this book, turns from the systemic argument to the personal one: the call to action for the executives building drone companies, the policymakers and congressional staff shaping the regulatory environment, the educators building the workforce pipeline, the investors allocating capital, the manufacturers closing the supply chain gap, and the community and regional leaders making economic development decisions that will shape their regions for a generation. The architecture does not build itself. It is built by people who decide to build it. Chapter 16 is for them.

Notes and Sources

1. CHIPS and Science Act of 2022, Public Law 117-167. Signed August 9, 2022. CHIPS Program Office established within the Department of Commerce. $52

billion in semiconductor manufacturing investment. Primary model for coordinated industrial strategy discussion throughout.

2. One Big Beautiful Bill Act (OBBBA), signed July 4, 2025. $1.4 billion appropriation for small UAS industrial base.

3. Defense Innovation Unit (DIU). Blue UAS Cleared List, updated August 2025. 23 cleared platforms, 14 framework components, 369 proposals from 19 nations in 2025 Refresh Challenge.

4. Federal Aviation Administration. UAS Collegiate Training Initiative Five-Year Review, 2020–2025. April 2025. 145+ institutions.

5. ASTM International. F3600 Standard for UAS Maintenance Technician.

6. UK Research and Innovation. Future Flight Challenge overview and program documentation, 2020–2026. £300 million government-industry joint program.

7. Australian Civil Aviation Safety Authority (CASA). Remotely Piloted Aircraft Systems (RPAS) and Advanced Air Mobility Strategic Regulatory Roadmap, 2023.

8. StartUs Insights. Global Drone Industry Report, 2025. More than 2.1 million workers worldwide in the drone industry.

9. FAA ASSURE Center of Excellence. ASSURE A83 2025 Annual Report, January 15, 2026. DJI platforms: 96%+ of detected U.S. national airspace operations.

10. The Next Web. "U.S. Banned DJI but China Controls 90% of Rare Earth Magnets and 99% of Drone Batteries Needed to Replace It." May 2026.

11. North Dakota Legislature, House Bill 1038, February 2025. $11 million Vantis-Pathfinder integration; $100 million cumulative state investment; $400 million leveraged federal and private capital.

12. Association for Uncrewed Vehicle Systems International (AUVSI). XPONENTIAL 2025 program documentation.

A Call to Action: Building What Comes Next

A Direct Address to the People Who Will Build the Uncrewed Aviation Era

"The moment does not wait for the institutions to catch up. The institutions have to decide to move. That decision, made by individual people in specific roles at a defined point in time, is how every era of American industrial leadership has actually begun."

— *Sharon Rossmark*

The Future We Can See, and the One We Must Choose

Return to the Tuesday from Chapter 1. The Assumption, Illinois spray mission. The Pittsburgh bridge inspection. The Phoenix delivery. The Texas border surveillance overnight. The Gulf of Mexico pipeline sweep. Recall the specific details: the operator watching a flight path populate on a tablet in the pre-dawn dark, the licensed remote pilot guiding thermal imagery through a utility corridor, the homeowner in her seventies who tracked the prescription delivery on her phone with the same casual familiarity she applies to a pizza order.

Now add five years. The Assumption operator is managing a fleet of 22 autonomous aircraft from that same tablet, a fleet whose combined coverage in a single morning exceeds what a dozen ground-based sprayers could accomplish in a week. The Pittsburgh inspection team is not guiding individual drones through a single corridor; it is analyzing AI-processed structural data from last week's 400-aircraft survey of the region's entire bridge inventory, a survey that took three days and cost less than a single helicopter inspection of a single bridge would have cost five years earlier. The Phoenix delivery corridor has been certified for routine BVLOS operations since 2027; the company running it operates 14 corridors in eight metropolitan areas and employs more logistics specialists, maintenance technicians, and operations managers in those markets than it employs pilots. The drones making these missions possible are built in

Indiana, assembled in Ohio, component-supplied from a precision manufacturing cluster in Texas that grew around a Blue UAS-qualified production contract and now employs 1,400 workers in a community whose previous anchor employer closed a decade ago.

This future is not a prediction. It is a projection from decisions that are being made, or not made, right now. The technology that makes it possible already exists. The regulatory framework that would enable it is being written. The workforce that would operate it is being trained, or not trained, in community colleges across the country where curriculum committees are deciding this semester whether drone technician credentials are worth building before the demand fully materializes. The manufacturing infrastructure that would produce the aircraft is being capitalized, or not capitalized, in boardrooms where the difference between a five-year commitment and a two-year pilot program determines whether a supply chain sovereignty bet becomes a structural advantage or a missed moment.

This book has spent fifteen chapters documenting the landscape, diagnosing the gaps, mapping the capital flows, arguing the strategic case, and building toward this chapter's purpose: a direct call to action, addressed to the people whose decisions will determine whether the possible future described above becomes the actual one. The call is not generic. The actions available to an aerospace CEO are different from those available to a community college president, which are different from those available to a congressional staff director, which are different from those available to a venture investor. This chapter speaks to each category directly.

The Synthesis: Technology Is Ready. The System Is Not.

The technology works. This is not a statement of faith. It is an observation grounded in the operational data of the Tuesday morning scenes in Chapter 1, scenes that are not promotional vignettes but documented, routine, unremarkable commercial operations that were science fiction fifteen years ago and are Tuesday now. The markets are real: the infrastructure inspection companies turning away contracts they cannot staff, the agricultural drone service providers running out of credentialed operators before running out of willing clients, the logistics firms that have built delivery corridor infrastructure ahead of the BVLOS framework that will make it commercially viable at scale. The economic case is validated by the investment data documented in Chapter 10.[2] The policy tools exist: Part 107, Remote ID, the BVLOS rulemaking, the Blue UAS framework, the OBBBA industrial base investment, the FAA CTI program, the WIOA workforce funding infrastructure.[6,7] None of these are sufficient on their own. Together, aligned by the coordinating architecture that Chapter 15 argued for, they are more than sufficient.

The technology is not waiting. The markets are not waiting. The capital is not waiting. The only thing that has not yet arrived at the scale the moment requires is the coordinated institutional will to deploy what already exists.

What is not yet built, not at the scale or coordination the moment requires, is the system surrounding the technology: the workforce that operates and maintains it at commercial density, the domestic manufacturing base that produces it with supply chain sovereignty, the investment architecture that funds the unglamorous but essential infrastructure that pure venture capital is not designed to fund, and the policy framework that governs it with the clarity and consistency that enterprise operators, institutional investors, and the global market require. These are not technological gaps. They are human choices. They are within reach. They require urgency. And they will be addressed, if they are addressed, by the specific people reading this chapter, making specific decisions in specific institutional contexts where their choices have more leverage than they may currently recognize.

What Each Sector Must Do: A Direct Call to Action

The call to action in this chapter is organized by role because the accountability for building the uncrewed aviation era is distributed across institutions and individuals whose specific leverage points are different. What follows is specific.

To the executives building drone companies and operating drone services: the workforce you need in three years does not exist yet, and the time to build the pipeline is now, before the need is acute enough to make hiring impossible. The companies that will own the next decade of commercial drone services are the ones building their workforce before they need it rather than recruiting against competitors for a pool that everyone is drawing from at once. Build internal training programs. Establish committed hiring relationships with the community colleges and technical schools in your operating regions.[3] Invest in technician training with the same seriousness that you invest in engineering hiring, because the maintenance technician who keeps your fleet airborne generates revenue as surely as the pilot who flies it. Stop treating workforce development as a cost center and start treating it as the capital investment it is: one whose returns are measured in operational capacity, client retention, and the competitive differentiation that comes from being the employer of choice in a market where qualified talent is constrained.

To the policymakers and congressional staff shaping the legislative and regulatory environment: the CHIPS Act demonstrated what coordinated industrial policy can accomplish when the strategic will exists to design it and sustain it.[5] Apply that model deliberately to uncrewed aviation. The domestic drone manufacturing base needs what the semiconductor manufacturing base needed: not just procurement restrictions that create demand, but the investment architecture that builds the supply. The workforce development system needs a credential framework with federal authority, a funding structure calibrated to actual employer needs, and the institutional patience to sustain investment through the years between program design and meaningful pipeline output. The BVLOS rulemaking needs finalization, implementation, and the regulatory certainty that allows the private sector to make the irreversible investments that commercial scale requires. The next FAA Reauthorization cycle is the most important near-term legislative opportunity in the industry's history.[6] Treat it accordingly.

To the educators running community colleges, technical schools, university aviation programs, private training organizations, and online education platforms: the workforce need is not waiting for the next curriculum review cycle. The operators turning away contracts documented in Chapter 3 are turning them away now, not in three years, and the credential pathways that would allow the professionals in your service area to fill those roles do not yet exist at the scale the market requires. This responsibility extends beyond traditional institutional programs: the private training companies and online platforms that reach career changers, working professionals, and rural learners who cannot attend campus-based programs are an essential part of the pipeline, and the quality, rigor, and employer alignment of their credentials matter as much as those of any accredited institution.[7] Move faster than your accreditation cycle has historically required, using the stackable credential model and the accelerated program design that the Unmanned Safety Institute's twelve-week maintenance technician program has demonstrated is achievable.[3] Partner with industry on equipment, curriculum alignment, and committed hiring agreements that transform your training program from a credential-production exercise into an actual employment pipeline. Recruit from the full geographic and professional range of your service area, the career-changer in an agricultural community, the veteran whose military UAS experience maps directly onto commercial requirements, the manufacturing worker whose precision skills are exactly what drone component production requires. These students are in your service area. They need a visible pathway and an honest economic pitch. Build both. One structural constraint on building it at scale deserves naming directly: workforce training pipelines cannot be built on enterprise-tier hardware. A training program that depends on platforms priced for defense procurement is a training program that cannot reach the students it most needs to reach. The absence of a domestically manufactured consumer and prosumer drone at a training-accessible price point is not a peripheral supply chain concern. It is a constraint on the pipeline's reach, and educators should be among the loudest voices making that case to the manufacturers and policymakers with the power to close the gap.

To the investors evaluating the drone and autonomy ecosystem: the systemic enablers are not charity investments. Workforce infrastructure, domestic component manufacturing, regional operator networks, vertiport development, and UTM integration are the preconditions for the returns on every other bet in the sector. The autonomy software platform whose revenue projections assume BVLOS operations at commercial scale requires a credentialed operator workforce. The domestic drone manufacturer whose defense contract pipeline justifies the Series C valuation requires a domestic supply chain that does not have a Chinese battery as its single point of failure.[11] The investor who allocates capital toward the enabling infrastructure, the patient equity, the industrial development finance, the long-hold-period investment in domestic component manufacturing, is not doing the sector a favor. They are capturing the compounding returns that come from owning the foundation on which everything else is built.

To the manufacturers building the domestic alternative: the consumer and prosumer market is not a secondary concern to be addressed after the defense and enterprise tiers are served. It is the foundation of the commercial drone services economy, and the policy actions that removed Chinese-manufactured platforms from the market did not redirect that economy. For the small business operator—the independent infrastructure inspector, the agricultural survey specialist, the real estate photographer running a sole proprietorship on Part 107 authority—the ban without a domestic alternative at a competitive price point is not a security solution. It is a market elimination. These operators built the commercial drone services industry. They trained the first generation of credentialed pilots, established the operational use cases that enterprise clients eventually adopted at scale, and demonstrated the economic viability of applications that are now the industry's core revenue drivers. Leaving them without an accessible, domestically sourced platform does not strengthen the American drone industry. It hollows out its commercial base while the defense and enterprise tiers are reinforced. The manufacturer that closes the consumer and prosumer price gap with a platform built to domestic sourcing standards will not be making a charitable concession to the small business segment. It will be capturing a market that is currently unserved, building the brand loyalty and operational ecosystem that enterprise market dominance follows, and providing the hardware on which the next generation of the workforce will be trained. That is not a secondary priority. It is the market opportunity that completes the domestic alternative.

To the community and regional leaders making economic development decisions: the drone economy is making location decisions right now, and the question is not whether it will arrive in your region but whether you will be positioned to capture its value when it does. Lead with workforce in your site selection pitch, not with tax incentives. Build the community college program before the manufacturer asks for it. Engage the FAA test site network before your competitors do.[9] Identify the anchor assets, the airport, the manufacturing facility, the agricultural sector, the infrastructure corridor, that give your region a specific operational argument for drone industry investment rather than a generic economic development appeal. The regions that are winning drone economy projects in 2026 are the ones that understood, two years ago, that the decision was being made before the announcement. The regions that will win in 2028 are the ones that understand that now.

The Capacity Argument: Why the Full Talent Pool Is a Strategic Imperative

There is a dimension of the workforce argument that this book has returned to repeatedly and that belongs in this final call to action stated as plainly as possible: the drone industry cannot fill its talent gap by recruiting from the same narrow pool that every other technology sector is already competing over. This is not a social observation. It is a capacity constraint, and ignoring it has exactly the same consequences as ignoring any other capacity constraint. The contracts go unfilled. The missions go unflown. The growth goes uncaptured.

The mid-career professional in a rural agricultural community who has spent years operating precision equipment and managing crop health data is a natural drone operator candidate. She is not applying, because no one has told her that the credential pathway exists and that the economic return on completing it is compelling. The veteran who spent four years operating Group 3 UAS platforms in a theater of operations has technical and operational experience that most commercial drone companies cannot produce in a training program, and he is driving a delivery truck because the military-to-commercial transition system has not built the bridge between his credentials and the commercial employer who needs exactly what he knows.[8] The manufacturing worker in a post-industrial community whose precision skills map directly onto drone component production is not in the hiring pipeline because the drone manufacturer has not established the recruiting presence in her community that would make the connection visible.

The industry cannot afford to leave trained, capable, motivated professionals on the sideline while complaining about a talent shortage. That is not a workforce problem. It is a recruiting problem, and recruiting problems have recruiting solutions.

Building the industry's workforce from the full depth of the available talent pool is the most strategically intelligent approach to building it at all. The communities closest to the industry's most economically significant applications, the agricultural heartland, the rural infrastructure corridors, the post-industrial manufacturing communities where precision skills are embedded in the workforce and idle factory space is available at below-market lease rates, are not peripheral to this industry's geography. They are central to it. And the individuals in those communities who become credentialed drone operators, maintenance technicians, fleet managers, and data analysts are not beneficiaries of the industry's generosity. They are its workforce.

The Window: Why the Next Two to Three Years Are Decisive

The BVLOS rulemaking finalization, expected in late 2026 or early 2027, is the single event most likely to accelerate the commercial drone market's trajectory.[2] When Part 108 takes effect, the infrastructure inspection companies, precision agriculture operators, and logistics providers who have been building BVLOS-ready operations ahead of the rule will have a structural advantage over competitors who have not. The workforce training programs that have been producing BVLOS-credentialed operators will have a market that did not previously exist at commercial scale. The preparation that seems premature today becomes the competitive foundation of 2028. The delay that seems prudent today becomes the catch-up cost of 2029.

The eVTOL certification milestones that Joby and Archer are targeting for 2026 and 2027 will, when achieved, release pent-up infrastructure investment in vertiport development, pilot training programs, and urban airspace management that is currently being held in reserve pending regulatory certainty.[10] The communities, real estate developers, utilities, and transit authorities that have positioned themselves ahead of those milestones will capture first-mover advantages in the infrastructure layer of a market whose value, at full commercial scale, is measured in hundreds of billions of dollars.

The international competitive pressure compounds both timelines. China has not retreated in response to NDAA restrictions and Blue UAS requirements. It has intensified its investment in the markets where those restrictions do not apply, deepened its control over the rare earth and battery supply chains that every drone manufacturer depends on, and continued developing the export market relationships that make DJI's global market position durable even as its U.S. market access narrows.[11] The United States currently leads in technology, in capital depth, in the breadth of its commercial ecosystem, and in the quality of its regulatory framework's ambition. The gap between its assets and the institutional coordination needed to convert them into durable competitive position is closable. It is not automatically closing.

The Closing Vision: Five Years From Today

Five years from today, if the decisions described in this chapter are made with the urgency they require, the uncrewed aviation industry will look substantially different from how it looks today. The BVLOS framework will be operational and normalized, its commercial impact already visible in the infrastructure inspection, precision agriculture, and logistics markets that have been constrained by the visual line of sight limitation for a decade. The domestic manufacturing base will have grown, not to the point of DJI-level market concentration, but to the point of genuine commercial viability, American-made platforms that compete on capability and cost, supported by domestic component suppliers who have built the scale to make the pricing competitive.[4] The workforce pipeline will be producing credentialed drone professionals at a rate that, while still not fully matching the pace of industry demand, is measurably closing the gap that has left contracts unfilled and missions unflown for years.

Five years from today, eVTOL operations will be running in at least a dozen American cities, carrying passengers on routes that are not yet the mass-market transportation alternative their proponents envision but that are demonstrably real, commercially viable, and building the operational track record on which the next phase of expansion depends. The workforce that serves these operations will have been built by programs that are being designed and funded right now, by educators who understood that the certification milestones were the signal to start building, not the signal to start planning.

This future is possible. It is not guaranteed. The only variable is whether the people with the power to build it decide to do so, now, with the urgency the moment requires, before the window closes on a decade that will not offer the same terms again.

The book you have been reading is an argument that this future is worth building and that the tools to build it exist. The technology is ready. The markets are real. The capital is available to those who design the structures that channel it toward the right places. The talent exists in deeper and more geographically distributed pools than the industry has yet learned to fish. The regulatory framework is being built by people who understand the stakes and are working faster than the history of comparable regulatory challenges would have predicted.

What remains is the decision. The decision made by the company executive who allocates this year's training budget as a strategic investment rather than a cost to minimize. The decision made by the community college president who assigns the drone technician program proposal to the fast track rather than the standard curriculum review cycle. The decision made by the congressional staffer who ensures that drone workforce development appears in the next FAA Reauthorization bill with the specificity and the funding that produces actual pipelines rather than aspirational language. The decision made by the investor who structures a fund around the enabling infrastructure rather than the headline platform. The decision made by the regional economic development director who leads with workforce in the site selection pitch and wins.

These decisions, made by individual people in specific roles at a defined point in time, are how every era of American industrial leadership has actually begun. Not with a single transformative legislative act or a single visionary company, but with the accumulated weight of thousands of institutional decisions, each one small enough to seem ordinary in the moment and large enough, in aggregate, to determine the outcome.

The airspace of tomorrow is being decided today. The question is not whether the uncrewed aviation era will arrive. It has arrived, eight minutes of flight at a time, in the fields and corridors and neighborhoods that Chapter 1 described on an unremarkable Tuesday morning. The question is whether the United States will lead it, or whether it will look back from a less advantaged position at the decisions it had the opportunity to make and the institutions it had the power to build.

The answer is not yet written. But the people writing it are identifiable, and many of them are reading this sentence right now.

Looking Ahead — Epilogue

The Epilogue that follows steps back from the argument to offer a personal reflection on what the uncrewed aviation era means, not just as an industrial opportunity or a policy challenge, but as a story about what America is capable of building when it decides to build deliberately.

Notes and Sources

1. Federal Aviation Administration. U.S. Civil Airmen Statistics, December 31, 2025. FAA Aeronautical Center, Oklahoma City, OK. Active Part 107 Remote Pilot Certificates: 492,311.

2. Federal Aviation Administration. Notice of Proposed Rulemaking: Part 108 BVLOS Operations. Published August 7, 2025. Implementation expected late 2026 – early 2027.

3. Unmanned Safety Institute (USI). UAS Maintenance Technician Program launch, February 2026. Twelve-week accelerated curriculum at Central Arizona College, Horry-Georgetown Technical College, and West Georgia Technical College.

4. Skydio, Inc. SkyForge Program Announcement, April 2026. $3.5 billion manufacturing expansion; 5,000+ projected jobs; $1 billion+ to domestic suppliers.

5. CHIPS and Science Act of 2022, Public Law 117-167. $52 billion semiconductor manufacturing investment. CHIPS Program Office model.

6. FAA Reauthorization Act of 2024, Public Law 118-63. Section 925 test site funding; BVLOS mandates; CTI program authorization.

7. FAA UAS Collegiate Training Initiative Five-Year Review, 2020–2025. April 2025. 145+ institutions.

8. Military Transition Toolkit. Army MOS 15W UAV Operator to Commercial Drone Pilot Career Pathway Guide, January 19, 2026.

9. Defense Innovation Unit (DIU). Blue UAS Cleared List, updated August 2025. 23 cleared platforms; 369 proposals from 19 nations in 2025 Refresh Challenge.

10. Joby Aviation and Archer Aviation. FAA Type Certification progress updates, 2025–2026.

11. The Next Web. "U.S. Banned DJI but China Controls 90% of Rare Earth Magnets and 99% of Drone Batteries Needed to Replace It." May 2026.

12. One Big Beautiful Bill Act (OBBBA), signed July 4, 2025. $1.4 billion for small UAS industrial base.

EPILOGUE

A Future We Choose

Somewhere in the country right now, a community college instructor is standing at the front of a classroom that did not exist two years ago, teaching a course that had no curriculum until she wrote one, using equipment donated by a company that had not yet identified the employer it would eventually become. The students in front of her are not the students anyone predicted. One of them spent twelve years in the Army operating Group 3 UAS platforms in conditions that no training simulator replicates. One of them drove a grain cart on her family's farm for a decade before she understood that the precision agriculture drone program at this institution was a credential pathway, not a hobbyist course. One of them is fifty-one years old and spent the morning before class troubleshooting a fleet management software problem that his employer's operations manual had not anticipated, which he solved because four months of serious technical training had given him the foundation to reason through problems the curriculum had not explicitly covered.

None of these students appear in the FAA's workforce projections. None of them were recruited by a national program. None of them followed a well-lit path. They found the path because someone decided to build it in their specific community, at their specific moment, for reasons that were practical and economic before they were anything else. That instructor, and the hundreds like her across the 145-plus institutions in the FAA's UAS Collegiate Training Initiative (CTI) network, represent something that no amount of federal investment or corporate strategy can substitute for: the distributed, local, human decision to build what the moment requires, made by people who understood the opportunity before the headlines confirmed it.

Throughout this book, the pattern has been consistent. Workforce shortages are not isolated challenges; they are structural constraints that shape what companies can build, how fast they can scale, and which opportunities they can pursue. Investment flows are robust, but not yet aligned with the long-term infrastructure the industry requires. Manufacturing capacity is growing, but not yet at the scale or diversity needed to support national resilience. These are not abstract concerns. They are the practical realities that determine whether an industry becomes globally competitive or remains perpetually dependent on systems built elsewhere.

And yet, the momentum is unmistakable. Communities are building training programs that did not exist a decade ago. Companies are integrating uncrewed systems into operations once considered too complex or too sensitive for automation. Policymakers are beginning to understand that aviation leadership in the 21st century will not be defined by a single aircraft or a single mission, but by an ecosystem, one that spans education, manufacturing, regulation, and workforce development in the kind of coordinated, sustained, patient investment that the Interstate Highway System required, that the semiconductor revival is requiring, and that this industry now requires in turn.

As this book goes to press, the White House released its America First Resilience Strategy (June 23, 2026), a document that places commercial drones explicitly alongside domestic manufacturing, artificial intelligence, and secure supply chains as technologies essential to national resilience.[1] The strategy's language is worth noting not for its policy novelty — it introduces no new FAA rules, procurement requirements, or funding programs — but for what it signals about how the administration understands the drone industry's strategic role. The document names the national goal plainly: to ensure that no adversary or hazard should be able to hold America or its core interests at risk, and it frames manufacturing agility, supply chain sovereignty, and trusted American technology as the instruments for achieving it.

For the commercial drone industry, the strategy's framing is clarifying in two specific respects. First, it confirms that the executive orders on domestic drone manufacturing, supply chain resilience, and trusted technology that have accumulated since June 2025 are not isolated procurement policy adjustments. They are expressions of a unified strategic posture, one that treats the ability to produce and deploy uncrewed systems at scale as a condition of national strength. Second, the strategy's four-tenet framework — prioritize risk, modernize systems, distribute responsibility, simplify governance — maps directly onto the institutional deficits this book has documented: workforce pipelines that reach only the most accessible talent pools, capital that flows toward hardware and away from the enabling infrastructure, manufacturing capacity that has grown at the finished-product level without yet addressing component sovereignty, and regulatory processes that have moved faster than most predicted but still carry the friction of frameworks not yet designed for the industry they govern.

There is one dimension the strategy does not address, and the omission is consequential. The policy architecture it confirms — NDAA restrictions, Blue UAS procurement frameworks, the FCC Covered List — was built for the defense and enterprise procurement tier. It has not yet produced a credible domestic alternative for the consumer and prosumer market: the affordable, accessible platforms on which small commercial operators, Part 107 pilots, STEM educators, and the community organizations building the talent pipeline all depend. The working drone operator who built a business around a DJI Mavic is not served by a platform priced for a government contract. The K–12 instructor whose students learn to fly on Chinese-made hardware does not have a domestic substitute at a price her school budget can reach. The National Resilience Strategy declares the direction. It does not resolve this gap, and neither, so far, has the broader policy response it ratifies. A manufacturing strategy that achieves sovereignty at the enterprise tier while leaving the commercial and educational market dependent on foreign-manufactured platforms has achieved compliance without capability.

The policy landscape around uncrewed aviation will continue to move, and the confusion that accompanies any rapidly evolving regulatory environment will continue with it. What the National Resilience Strategy does is fix a coordinate: whatever rules change, whatever programs come and go, the strategic direction is declared. Drones are infrastructure. Workforce is readiness. Manufacturing is sovereignty. The question the strategy does not answer — the one no document from Washington can answer — is whether the people with the power to build what the moment requires will choose to do so.

The next decade will be decisive. Not because the technology will suddenly leap forward — it has already leaped far enough — but because the institutions around it will either rise to meet the moment or fall short of it. Leadership in uncrewed aviation will not belong to the nation with the most patents or the most prototypes. It will belong to the nation that builds the talent pipelines, the industrial base, and the strategic alignment required to turn demonstrated capability into sustained competitive advantage.

The opportunity is real. The demand signals are clear. The stakes are high. The instructor in that classroom, the veteran learning to translate military operational discipline into commercial mission management, the farmer's daughter who recognized in a community college course the beginning of a career the agriculture industry will depend on, each of them represents a bet, made by a specific institution on a specific day, that the future described in this book is worth building toward. They are right. And the only thing that determines whether those individual bets compound into national capability is the collective decision, made by executives, policymakers, investors, educators, and community leaders, to make the same bet, at the scale the moment requires.

"The question now is whether we have the will to build what the future requires."

Sharon Rossmark

Notes and Sources

1. The White House. President Trump's America First Resilience Strategy. June 23, 2026. Washington, D.C.: The White House. Primary document: https://www.whitehouse.gov/wp-content/uploads/2026/06/2026-National-Resilience-Strategy-1.pdf. Official release: https://www.whitehouse.gov/releases/2026/06/white-house-unveils-president-trumps-america-first-resilience-strategy/. The strategy explicitly names the commercial drone industry as contributing to national resilience alongside domestic manufacturing, artificial intelligence, and secure supply chains, and establishes as the National Resilience Goal that "no adversary or hazard should be able to hold America or our core interests at risk."

About the Author

Sharon Rossmark is CEO and Founder of Women and Drones and the author of Future Flight. She has spent more than a decade building workforce programs, advising companies, and working alongside the educators, operators, and policymakers constructing the American drone industry from the inside.

www.ingramcontent.com/pod-product-compliance
Lightning Source LLC
LaVergne TN
LVHW022027051126
851858LV00019B/310/J